AF522239

TEXT BOOK
OF
ELEMENTARY STATISTICS

DPH MATHEMATICS SERIES

TEXT BOOK
OF
ELEMENTARY STATISTICS

By

A.K. Sharma

DISCOVERY PUBLISHING HOUSE
NEW DELHI-110002

First Published - 2005

Reprinted - 2017

ISBN: 978-81-7141-953-1

Text Book of Elementary Statistics

Published by:

DISCOVERY PUBLISHING HOUSE PVT. LTD.
4383/4B, Ansari Road, Darya Ganj
New Delhi-110 002 (India)
Phone: +91-11-23279245, 43596064-65
Fax: +91-11-23253475
E-mail: discoverypublishinghouse@gmail.com
sales@discoverypublishinggroup.com
web: www.discoverypublishinggroup.com

Printed at:
Infinity Imaging Systems
Delhi

Preface

This book Elementary statistics has been written the requirement of graduate students of Indian universities. This book is also helpful for the candidate appearing in various competitive examination. The present publication has been written to present a logical and practical approach to the various complexities and intricacies of the subject at length inspite of comprehensiveness.

We have tried to our best to keep the book free from the misprints. The author shall be grateful to the rears who point out errors and omissions which inspite of all care, might have been there.

The author in general hope that the present book will be warmly received by the students and teachers. We shall indeed be very thankful to our colleagues for their recommending this book for their students.

A.K. Sharma

Contents

1

CLASSIFICATION AND TABULATION OF DATA

INTRODUCTION

The phrase "classification and tabulation" has been used, classification is, in effect, only the first step in tabulation, for, in general, items having common characteristics must be brought together before the data can be displayed in tabular form.

MEANING OF CLASSIFICATION

Sorting facts on one basis of classification and then on another basis is called cross classification. This process can be repeated as many times as there are possible bases of classification. Classification of data is a function very similar to that of sorting letters in a post-office. It is well known that the letters collected in a post-office are sorted into different lots on a geographical basis.

Objects of Classifications

The principal objective of classification data are:

1. To pinpoint the most significant features of the data at a glance.
2. To give prominence to the important information gathered while dropping out the unnecessary elements.
3. To enable a statistical treatment of the material collected.
4. To condense the mass of data in such a manner that similarities and dissimilarities can be readily apprehended. Millions of figures can thus be arranged in a few classes having common features.
5. To facilitate comparison.

Types of Classification

The data can be classified on the following four basis :

1. Qualitative, *i.e.,* according to some attributes.

2. Quantitative, *i.e.,* in terms or magnitudes.
3. Geographical, *i.e.,* area-wise, *e.g.,* cities, districts, etc.
4. Chronological, *i.e.,* on the basis of time.

1. Qualitative Classification

In qualitative classification data are classified on the basis of some attribute or quality such as sex, color of hair, literacy, religion, etc.

The type of classification where only two classes are formed is also called two-fold or dichotomous classification. If instead of forming only two classes we further divide the data on the basis of some attribute or attributes so as to form several classes, the classification is known as manifold classification.

2. Quantitative Classification

Quantitative classification refers to the classification of data according to some characteristics that can be measured, such as height, weight, etc. For Ex. the students of a college may be classified according to weight as follows:

Weight (in lb.s)	No. of Students
90–100	100
100–110	220
100–120	250
120–130	370
130–140	80
140–150	60
Total	**1,080**

Such a distribution is known as empirical frequency distribution or simple frequency distribution.

The following are two Ex.s of discrete and continuous frequency distributions :

No. of Children	No. of Families	Weight (lb.s)	No. of persons
0	15	100–110	15
1	20	110–120	20
3	70	120–130	50
4	120	130–140	55
5	270	140–150	15
6	70	150–160	5
Total	575	Total	160

(a) Discrete Frequency Distribution. (b) Continuous Frequency Distribution.

3. Geographical Classification

In this type of classification data are classified on the basis of geographical or locational differences between the various items. For instance, the production of sugarcane in India may be presented State-wise in the following manner:

Production of Sugarcane for the Year 1976 (Figures imaginary)

Name of State	Sugarcane Production (in million tonnes)
Uttar Pradesh	68
Bihar	28
Tamil Nadu	18
Maharashtra	14
Other States	12
Total	140

4. Chronological Classification

When data are observed over a period of time the type of classification is known as chronological classification. For Ex., we may present the figures of population (or production, sales, etc.) as follows :

Population of India from 1921 to 1971

Year	Production of India (in millions tonnes)	Year	Production of India (in millions tonnes)
1921	348	1951	538
1931	376	1961	576
1941	413	1971	678

FORMATION OF A DISCRETE FREQUENCY DISTRIBUTION

In the formation of discrete frequency distribution, we have just to count the number of times a particular value is repeated which is called the frequency of that class. In order to facilitate counting, prepare a column of "tallies". In another column, place all possible values of variables from the lowest to the highest. Then, put a bar (vertical line) opposite the particular value to which it relates. To facilitate counting, blocks of five bars are prepared and some space is left in between each block. We finally count the number of bars corresponding to each value of the variable and place it in the column entitled 'frequency'.

Example:

In a survey of 35 families in a village, the number of children per family was recorded and the following data obtained :

1	*3*	*2*	*3*	*4*	*5*	*6*
7	*2*	*3*	*4*	*5*	*2*	*5*
8	*4*	*5*	*10*	*6*	*3*	*2*
7	*6*	*5*	*4*	*3*	*7*	*8*
9	*1*	*9*	*4*	*5*	*4*	*3*

Solution:

Frequency Distribution of the Number of Children

No. of Children	*Tallies*	*Frequency*
1	|	1
2	||||	4
3	~~||||~~ |	6
4	~~||||~~ |	6
5	~~||||~~	6
6	|||	3
7	|||	3
8	||	2
9	|	1
10	||	2
		Total 35

FORMATION OF CONTINUOUS FREQUENCY DISTRIBUTION

The following technical terms are important when a continuous frequency distribution is found.

(i) **Class Limits :** The class limits are the lowest and the highest values that can be included in the class. For Ex., take the class 10-20. The lowest value of the class is 10 and the highest 20. The two boundaries of a class are known as the lower limit and the upper limit of the class.

(ii) **Class Intervals :** The spare of a class, that is the difference between the Upper and Lower Limit is known as class interval for Ex., in the class 50-100, the class interval is 100 (*i.e.,* 100 minus 50). An important decision while constructing a frequency distribution is about the width of the class interval. A simple formula to obtain the estimate of appropriate class interval *i.e., i* is.

$$i = \frac{L - S}{k}$$

where, L = largest item,

S = smallest item,

k = the number of classes

(iii) **Class Frequency :** The number of observations corresponding to a particular class i known as the frequency of that class or the class frequency. In the following Ex., the frequency of the class 0-50 is 50 which implies that there are 50 persons having income between Rs. 0-50.

Class Mid-point or Class Mark : It is the value lying half-way between the lower and upper class limits of a class-interval, Mid-point of a class is ascertained as follows :

$$\text{Mid-point of a class} = \frac{\text{Upper limit of the class} - \text{Lower limit of the class}}{2}$$

There are two methods of classifying the data according to class-intervals, namely:

(i) 'Executive' method, and

(ii) 'Inclusive' method.

(i) **'Exclusive' Method :** When the class intervals are so fixed that the upper limit of one class is the lower limit of the next class it is known as the 'exclusive' method of classification. The following data are classified on this basis :

Income (Rs.) per day	*No. of Persons*
0–50	50
50–100	100
100–150	200
150–200	150
200–250	40
250–300	10
	Total 550

It is clear that the 'exclusive' method ensures continuity of data inasmuch as the upper limit of one class is that lower limit of the next class. Thus, in the above Ex., there are 50 persons whose income is between Rs. 0-49.99. A person whose income is Rs. 50 would be included in the class Rs. 50-100.

A better way of expressing the classes when exclusive method is followed is :

Income (Rs.)	*No. of Persons*
0 but under 50	50
50 but under 100	100
100 but under 150	200
150 but under 200	150
200 but under 250	40
250 but under 300	10
	Total 550

(ii) **'Inclusive' Method :** Under the 'inclusive' method of classification, the upper limit of one class it included in that class itself. The following Ex. illustrates the method :

Income (Rs.)	*No. of Persons*
0–49	50
50–99	100
100–149	200
150–199	150
200–249	50
250–299	10
	Total 560

In the class 100-149 we include persons whose income is between Rs. 100 and Rs. 149. It the income of a person is exactly Rs. 150 in the next class.

TABULATION OF DATA

A table is a systematic arrangement of statistical data in columns and rows. Rows are horizontal arrangements whereas columns are vertical ones. The purpose of a table is to simplify the presentation and to facilitate comparisons. The simplification results from the clear-cut and systematic arrangement, which enables the reader to quickly locate desired information. Comparison is facilitated bringing related items of information close together.

Role of Tabulation

Tables make it possible for the analyst to present a huge mass of data in a detailed orderly manner within a minimum of space. Because of this, tabular presentation is the cornerstone of statistical reporting. The significance of tabulation will be clear from the following points:

1. *It simplifies complex data :* When data are tabulated all unnecessary details and repetitions are avoided. Data are presented systematically in columns and rows. Hence, the reader gets a very clear idea of what

the table presents. There is thus a considerable saving in time taken in understanding what is represented by the data and all confusion is avoided. Also a large amount of space is saved because of non-duplicating of his headings and designations; the description at the top of a column serves for all the terms beneath it.

2. *It facilitates comparison :* Tabulation facilitates comparison. Since a table is divided into various parts and for each part there are totals and sub-totals, the relationship between different parts of data can be studied much more easily with the help of a table than without it.
3. *It reveals patterns :* Tabulation reveals patterns within the figures which cannot be seen in the narrative form. It also facilitates the summation of the figures if the reader desires to check the totals.
4. *It gives identity to the data :* When the data are arranged in a table with a title and number they can be distinctly identified and can be used as a source reference in the interpretation of a problem.

PARTS OF A TABLE

The main parts of a table in general are following :

1. Table number
2. Title of the table
3. Caption
4. Stub
5. Body of the table
6. Headnote
7. Footnote

1. **Table Number :** Each table should be numbered. There are different practices with regard to the place where this number is to be given. The number may be given either in the centre at the top above the title or Inside of the title at the top or in the bottom of the table on the left-hand side. However if space permits the table number should be given in the, centre as is shown in the specimen table given on page. Where there are many columns, it is also desirable to number each column so that easy reference to it is possible.
2. **Title of the Table :** Every table must be given suitable title. The title is a description of the contents of the table. A complete title has to answer the questions what, where and when in that sequence. In other words:

   (i) what precisely are the data in the table (*i.e.*, what categories of statistical data are shown)?

   (ii) when the data occurred (*i.e.,* the specific time or period covered by the statistical materials in the table)?

(iii) where the data occurred (*i.e.*, the precise geographical, political or physical area covered)?

The title should be clear, brief and self-explanatory. However, clarity should not be sacrificed for the sake of brevity. Long titles cannot be read as promptly as short titles, but at times they may have to be used for the sake of clarity. The title should be so worded that it permits one and only one interpretation. It should be in the form of a series of phrases rather than complete sentences. Its lettering should be the most prominent of any lettering on the table.

3. **Caption :** Caption refers to the column headings. It explains, what the column represents. It may consist of one or more column headings. Under a column heading there may be sub-heads. The caption should be clearly defined and placed at the middle of the column. If the different columns are expressed in different units, the units should be mentioned with the captions. As compared with the main part of the table the caption should be shown in smaller letters. This helps in saving space.
4. **Stub :** As distinguished from caption, stubs are the designations of the rows or row headings. They are at 'the extreme left and perform the same function for the horizontal rows of numbers in the table as the column headings do for the vertical columns of numbers. The stubs are usually wide than column headings but should be kept as narrow as possible without sacrificing precision and clarity of statements.
5. **Body :** The body of the table contains the numerical information. This is the most vital part of the table. Data presented in the body arranged according to description are classifications of the captions and stubs.
6. **Headnote :** It is a brief explanatory statement applying to all or a major part of the material in the table, and is placed below the point centered and enclosed in brackets. It is used to explain certain points relating to the whole table that have not been included in the title nor in the captions or stubs. For example, the unit of measurement is frequently written as a headnote, such as "in thousands" or "in million tonnes" or "in crores", etc.
7. **Footnotes :** Anything in a table which the reader may find difficult to understand from the title, captions and stubs should be explained in footnotes. If footnotes are needed they are placed directly below the body of the table. Footnotes are used for the following main purposes:

(i) To clarify anything in the table.

(ii) To point out any exceptions as to the basis of arriving at the data, for example, sales recorded at 'ex-factory price' for some of the entries and at 'delivered price" for others. Any heterogeneity in the data recorded must be disclosed to avoid wrong conclusions.

(iii) Any special circumstances affecting the data, for example, strike, lock-out, fire, etc.,

(iv) To give the source in case of secondary data. The reference to the source should be complete in itself. For example, if the data is obtained from some periodical, its name, date of publication, page number, table number, etc., should be mentioned so that if the user wishes to check the data from the original source or considers later data from the same source he will know where to look for the information.

GENERAL RULES OF TABULATION

"In collection and tabulation common sense is the chief requisite and experience the chief teacher." However, the following general considerations may be kept in view while tabulating data:

1. The table should suit the size of the paper usually with more rows than columns. In making a suitable layout it may be necessary to alter the original design. The alteration often consists In changing the rows to columns or the other way round. For this reason, it Is desirable to make a rough draft of the table before the figures are entered In it. Space must be allowed for reference or any other matter which is to be Included in the table.
2. The table should not be overloaded with details. If many characteristics are to be shown it is not necessary to load them all in one table; rather a number of tables should be prepared, each table complete in itself and serving a particular purpose.
3. Indicate a zero quantity by a zero, and do not use zero to indicate information which is not available. If it is not available, show this fact by the letter N.A. or dash (—).
4. A column entitled 'miscellaneous' should be added for data which do not fit in the classification made.
5. Be explicit. The expression "etc." is bad form in a table, since the reader may not readily discover what it refers to. In fact clarity is the most Important feature of tabular presentation of any kind of statistical data.

6. The arrangement of the table should be logical and Items related to each other should be placed nearabout and. If possible. In the same group. Derivative figures such as totals, averages and percentages should be placed near the original figures. Columns and rows should be numbered for identification since reference is more easily given by quoting numbers than the title of the column.
7. Abbreviations should be avoided especially in titles and headings. For example "yr'' should not be used for 'year'.
8. In all tables the captions and stubs should be arranged in some systematic order. It would make the table easier to read and allow more important items to be emphasised. The arrangement of items basically depends upon the type of data. However, the principal bases for arranging items are the following:
   (a) Alphabetical, *i.e.,* arrangement according to alphabets. The type of arrangement is very common in general purpose or reference tables.
   (b) Chronological, *i.e.,* arrangement according to time. This type of arrangement is of particular value in presenting historical data.
   (c) Geographical, *i.e.,* arrangement of data in certain territorial units such as countries, cities, districts, etc.
   (d) Conventional, or arrangement in a customary order such as men, women and children or Hindus, Muslims, Sikhs and Christians.
   (e) Items may be arranged according to size, *i.e.,* the numerical importance of the items, the largest items being given first and the smallest in the last. This arrangement may be reversed, if necessary.
9. Where standard classifications have been prepared it is usually desirable to employ them, as they are superior to hastily constructed individual classification.
10. If certain figures are to be emphasized they should be in distinctive type or in a 'box' or circle or between thick lines.
11. Percentage and ratios should be computed and shown, if necessary. Frequently, figures in tables become more meaningful if expressed as percentages or (less often) as ratios. In constructing a table, therefore, it is important to decide whether or not it can be Improved in this way. If it can, additional column should be Inserted in the table and the percentages (or ratios) computed and entered. Such percentages and ratios are sometimes called derived statistics.

12. Figures should be rounded off to avoid unnecessary details in the table and a footnote to this effect should be given. For example, the figures may be taken to the nearest rupees and paise be eliminated.
13. The point of measurement should be clearly defined and given in the table such as income in rupees or weight in pounds, etc.

CONSIDERATIONS IN THE CONSTRUCTION OF FREQUENCY DISTRIBUTIONS

It is difficult to lay down any hard and fast rules for constructing a frequency distribution much depends on the nature of the given data and the object of classification.

However, the following general considerations may be borne in mind for ensuring meaningful classification of data :

1. The starting point, *i.e.,* the lower limit of the first class, should either be zero or 5 or multiple of 5.
2. The number of classes should preferably be between 5 to 20. However, there is no rigidity about it. The classes can be more than 20 depending upon the total number of items in the series and the details required, but they should not be less than five because in that case the classification may not reveal the essential characteristics. The choice of number of classes basically depends upon:
   (a) the number of figures to be classified,
   (b) the magnitude of the figures,
   (c) the details required, and
   (d) case of calculation of further statistical work.
3. As far as possible one should avoid such values of class-intervals, as 3, 7, 11, 26, 39, etc. Preferably, one should have class-intervals of either five or multiples of 5 like 10, 20, 25, 100 etc. The reason is that the human mind is accustomed more to think in terms of certain multiples of 5, 10 and the like. However, where the data necessitate a class-intervals of less than 5 it can be any value between 1 and 4.
4. The ensure continuity and to get correct class-interval we should, adopt 'exclusive' method of classification. However, were 'inclusive' method has been adopted it is necessary to make an adjustment to determine the correct class-interval and to have continuity. The adjustment consists of finding the difference between the lower limit of the second class and the upper limit of the first class, dividing the difference by two, subtracting the value so obtained from all lower

limits and adding the value to all upper limits. This can be expressed in the form of a formula as follows :

$$\text{Correction factor} = \frac{\text{Lower limit of the 2nd class} - \text{Upper limit of the 1st class}}{2}$$

How the adjustment is made when data are given by inclusive method can be seen from the following Ex.s:

Weekly Wages (in Rs.)	No. of Workers	Weekly Wages (in Rs.)	No. of Workers
10–19	5	40–49	8
20–29	10	50–59	2
30–39	15		

To adjust the class limits, we take here the difference between 20 and 19, which is one. By dividing it by two we get ½ or 0.5. This (0.5) is called the correction factor. Deduct 0.5 from the lower limits of all classes and add 0.5 to upper limits. The adjusted classes would then be as follows :

Weekly Wages (in Rs.)	No. of Workers	Weekly Wages (in Rs.)	No. of Workers
9.5–19.5	5	39.5–49.5	8
19.5–29.5	10	49.5–59.5	2
29.5–39.5	15		

It should be noted that before adjustment the class-intervals was 9 but after adjustment, it is 10. Observe another case.

Variable	Frequency
5–9.5	8
10–14.5	10
15–19.5	2

The correction factor here is $\frac{(10 - 9.5)}{2} = 0.25.$

After adjustment the classes will be :

The class-interval now is 5 and not 4.5. Taking a third Ex., if the class limits are :

Variable	Frequency
5–9.99	8
10–14.99	10
15–19.99	2

The correction factor would be $\frac{(10 - 9.99)}{2} = \frac{0.01}{2} = 0.005$.

After adjustment the classes will become :

Variable	Frequency
4.995–9.995	8
9.995–14.995	10
14.995–19.995	2

5. Wherever possible, it is desirable to use class intervals of equal sizes because comparisons of frequencies among classes are facilitated and subsequent calculations from the distribution are simplified. However, this is not always a practical procedure.

Open-end distribution presents problems of graphing and further analysis. When the frequency distribution is being employed as the only technique of presentation, open-end classes do not seriously reduce its usefulness as long as only a few items fall in these classes. However, use of the distribution for purposes of further mathematical computation is difficult because a mid-point value, which can be used to present the class, cannot be determined for an open-end class.

6. In any frequency distribution the sizes of items or the values are indicated on the left-hand side and the number of times the items in those sizes or values have repeated are indicated by frequencies on the right-hand side opposite to the respective sizes or values.

SOLVED EXAMPLES

Example 1:

Point out the mistakes in the following table drawn to show the distribution of population, according to sex, age and literacy :

Sex.	*0 to 25*	*25 to 50*	*50 to 75*	*75 to 100*
Males	—	—	—	—
Females	—	—	—	—

Solution:

All the characteristics are not revealed in the above table; the characteristic of literacy has been completely ignored. Even otherwise the table need to be re-arranged as follows :

Table Showing the Distribution of Population According to Age, Sex and Literacy

	Literates			Illiterates			Total		
Age group	M	F	Total	M	F	Total	M	F	Total
0 to 25	—	—	—	—	—	—	—	—	—
25 to 50	—	—	—	—	—	—	—	—	—
50 to 75	—	—	—	—	—	—	—	—	—
75 to 100	—	—	—	—	—	—	—	—	—
Total	—	—	—	—	—	—	—	—	—

Example 2:

Draft a blank table to show the distribution of personnel in a manufacturing concern according to :

(a) Sex : males and females.

(b) Three grades of salary : below Rs. 300, Rs. 300–500, Rs. 500 and above.

(c) Two period : 1978 and 1979

(d) Three age-groups : below 25, 25 and under 40, 40 and over.

Solution:

Table Showing Distribution of Personnel According to Sex, Salary and Age-Groups for Two Years

		Salary Grade											
Year	Age	Below Rs. 300			Rs. 300–500			Rs. 500 & above			Total		
	Groups	M	F	Total	M	F	Total	M	F	Total	M	F	Total
	Below 25	—	—	—	—	—	—	—	—	—	—	—	—
1978	25 and under 40	—	—	—	—	—	—	—	—	—	—	—	—
	40 and above	—	—	—	—	—	—	—	—	—	—	—	—
	Total	—	—	—	—	—	—	—	—	—	—	—	—

	Below 25	—	—	—	—	—	—	—	—	—	—	—	—
1979	25 and under 40	—	—	—	—	—	—	—	—	—	—	—	—
	40 and above	—	—	—	—	—	—	—	—	—	—	—	—
	Total	—	—	—	—	—	—	—	—	—	—	—	—

Example 3:

Following figures give the ages of newly married husbands and their wives in years. Represent the data by a frequency distribution.

Age of husband	*Age of Wife*	*Age of husband*	*Age of Wife*
24	*17*	*25*	*17*
26	*18*	*26*	*18*
27	*19*	*27*	*19*
25	*17*	*25*	*19*
28	*20*	*27*	*20*
24	*18*	*26*	*19*
27	*18*	*25*	*17*
28	*19*	*26*	*20*
25	*18*	*26*	*17*
26	*19*	*26*	*18*

Solution:

Frequency Distribution of the Age of Husbands and Wives

	Age of husbands					**Total**
Age of wives	**24**	**25**	**26**	**27**	**28**	
17	I (1)	III (3)	I (1)	—	—	5
18	I (1)	I (1)	III (3)	I (1)	—	6
19	—	I (1)	II (2)	II (2)	I (1)	6
20	—	—	I (1)	I (1)	I (1)	3
Total	2	5	7	4	2	20

Example 4:

Out of a total number of 1,807 women who were interviewed for employment in a textile factory of Bombay, 512 were from textile areas and the rest from the non-textile areas. Amongst the married women who belonged to textile areas, 247 were experienced and 73 inexperienced, while for non-

textile areas, the corresponding figures were 49 and 520. The total number of inexperienced women 918 were unmarried, and of these the number of experienced women in the textile and non-textile areas was 154 and 16 respectively. Tabulate.

Solution:

Table Showing the Marital Status of 1,807 Women Residing in Textile and Non-Textile Areas

	Textile Areas			Non-textile Areas			Total		
	M	**U**	**Total**	**M**	**U**	**Total**	**M**	**U**	**Total**
Experienced	247	154	401	49	16	65	296	170	466
Inexperienced	73	38	111	520	710	1,230	593	748	1,341
Total	320	192	512	567	726	1,295	889	918	1,807

Example 5:

In a sample study about coffee habit in two tons, the following information were received :

Town A *Females were 40%; Total coffee drinkers were 45% and Male non-coffee drinkers were 20%.*

Town B *Males were 55%; Males non-coffee drinkers were 30% and Females coffee drinkers were 15%.*

Represent the above data in a tabular form.

Solution:

Table Showing the Coffee Drinking Habit of Town A & B

(in percentage)

Attribute	Town A			Town B		
	Males	**Females**	**Total**	**Males**	**Females**	**Total**
Coffee Drinkers	40	5(a)	45	25c)	15	40(e)
Non-coffee Drinkers	20	35(b)	55	30	30(d)	60(f)
Total	60	40	100	55	45	100

Figures (a), (b), (c), (d), (e) and (f), are obtained by simple process of deduction after taking into consideration the given information.

Example 6:

Represent the following information in suitable tabular form with proper rulings and headings :

The Annual Report of the Ishapore Public Library reveals the following points regarding the reading habits of its members :

Out of the total 3,713 book issued to the members in the month of June 1970, 2,100 were fictions. There were 467 members of the library during the period and they were classified into five classes A, B, C, D, and E. The number of members belonging to the first four class were respectively 15, 176, 98 and 129; and the number of fictions issued to them were 103, 1,187, 647 and 58 respectively. Number of books, other than textbooks and fictions, issued to these four classes of members were respectively 4,390, 217 and 341. The textbooks were issued only to members belonging to he classes C, D and E and the number of textbooks issued to them were respectively t3, 317 and 160.

During the same period, 1,246 periodicals were issued. Thee included 396 technical journals of which 36 were issued to members of class B, 45 to class D and 315 to class E.

To members of the classes B, C, D and E the number of other journals issued were 419, 26, 231 and 99 respectively.

The report, however, showed an increased by 3.9% in the number of books issued over last month, though was a corresponding decrease by 6.1% in the number of periodicals and journals issued to members.

Solution:

Table Showing the Reading Habits of People

		Numbers of books issued				Numbers of periodicals issued		
Class of members	**No. of members in each class**	**Fiction**	**Text books**	**Other than fiction & text-books**	**Total**	**Tech-nical**	**Others**	**Total**
A	15	103	—	4	107	—	75	75
B	176	1,187	—	390	1,577	36	419	455
C	98	647	3	217	867	—	26	26
D	129	58	317	341	716	45	231	276
E	49	105	160	181	446	315	99	414
Total	467	1,200	480	1,133	3,713	396	850	1,246
%age increase (+) decrease (–)	—	—	—	—	+ 3.9	—	—	– 6.1
May, 1970								

The main principles of classification adopted are :

(i) One major group is the class of members.

(ii) Among reading matter, the two major classifications are—Books and Periodicals.

(iii) The sub-groups under each of the two groups under (ii) above are (a) fiction, (b) text books, and (c) others under books and (a) technical, and (b) other under periodicals.

(iv) The number of members belonging to class 'E' is derived by deducting the sum of the members (given) under the four classes A, B, C and D from the total number of members.

Example 7:

A sample consists of 34 observations recorded correct to the nearest integer, ranging in value from 201 to 337. If it is decided to use seven classes of width 20 integers and to begin the first class at 199.5, find the class limits and class marks of the seven classes.

Solution:

Since it is decide to begin with 199.5 and take a class interval of 20, the first class be 199.5–219.5, the second 219.5–239.5 and so on. The class mark shall be obtained by adding the lower and upper limits and dividing it by 2. Thus for the first class, the class mark shall be (199.5 + 21.5)/2 = 209.5. Since class interval is equal throughout the other class marks can be obtained simply by adding 20 to the preceding class mark. The following table gives the class limits and class marks of the seven classes :

Class limits	Class marks
199.5–219.5	209.5
219.5—239.5	229.5
239.5–259.5	249.5
259.5–279.5	269.5
279.5–299.5	289.5
299.5–319.5	309.5
319.5–339.5	329.5

Example 8:

Present the following information in a suitable tabular form:

In 1965 out of a total of 1,750 workers of a factory, 1,200 were members of a trade union.

The number of women employed was 200, of which 175 did not belong to a trade union. In 1970 the number of union workers increased to 1,580 of which 1,290 were men. On the other hand, the number of non-union workers fell down to 208, of which 180 were men.

In 1975, there were 1,800 employees who belonged to a trade union and 50 who did not belong to a trade union. Of all the employees in 1970, 300 were women of whom only 8 did not belong to a trade union.

Solution:

Table Showing the Sex-wise Distribution of Union and Non-union Members for 1965, 1970 and 1975

Category	1965			1970			1975		
	M	F	Total	M	F	Total	M	F	Total
Members	1,175	25	1,200	1,290	290	1,580	1,508	292	1,800
Non-members	375	175	550	180	28	208	42	8	50
Total	1,550	200	1,750	1,470	318	1,788	1,550	300	1,850

Example 9:

In a trip organised by a college there were 80 persons, each of whom paid Rs. 15.50 on an average. There were 60 students each of whom paid Rs. 16. Members of the teaching staff were charged at a higher rate. The number of servants was 6 (all males) and they were not charged anything. The number of ladies was 20% of the total of which one was a lady staff member.

Solution:

Table Showing the Type of Participants, Sex and Contribution Made

Types Participants	Sex			Contribution per member (Rs.)	Total Contribution (Rs.)
	Males	Females	Total		
Students	45	15	60	16.00	960
Teaching Staff	13	1	14	20.00	280
Servants	6	—	6	—	—
Total	64	16	80	—	1,400

Notes: 1. Total contribution = Average contribution × No. of person who joined the trip

$= 15.5 \times 80 = 1,240$

2. Contribution of the staff per head has been obtained by deducting the contribution of students from the total dividing the difference by the number of teaching staff, *i.e.*,

$$\frac{(1240-(60\times 16)}{14}=\frac{1240-960}{14}=\frac{280}{14}=\text{Rs.}20\,.$$

Example 10:

Draw a blank table to present the information regarding the college students according to :

(a) Faculty — Art, Commerce, Science.

(b) Class — Degree and Pre-University Class.

(c) Sex — Male and Female.

(d) Age — below 20, above 20,

(e) For 2 years — 1978 and 1979.

Solution:

Table Showing the Sex-wise Distribution of the People of Different Religions, Localities and Age-groups

		Degree Classes								
		Males			**Females**			**Total**		
Years	**Faculties**	**Below 20**	**Above 20**	**Total**	**Below 20**	**Above 20**	**Total**	**M**	**F**	**Total**
1978	Art	—	—	—	—	—	—	—	—	—
	Commerce	—	—	—	—	—	—	—	—	—
	Science	—	—	—	—	—	—	—	—	—
	Total	—	—	—	—	—	—	—	—	—
1979	Art	—	—	—	—	—	—	—	—	—
	Commerce	—	—	—	—	—	—	—	—	—
	Science	—	—	—	—	—	—	—	—	—
	Total	—	—	—	—	—	—	—	—	—

		Pre-University Classes								
		Males			**Females**			**Total**		
Years	**Faculties**	**Below 20**	**Above 20**	**Total**	**Below 20**	**Above 20**	**Total**	**M**	**F**	**Total**
1978	Art	—	—	—	—	—	—	—	—	—
	Commerce	—	—	—	—	—	—	—	—	—
	Science	—	—	—	—	—	—	—	—	—
	Total	—	—	—	—	—	—	—	—	—

1979	Art	—	—	—	—	—	—	—	—	—
	Commerce	—	—	—	—	—	—	—	—	—
	Science	—	—	—	—	—	—	—	—	—
	Total	—	—	—	—	—	—	—	—	—

M = Males, F = Females

Hand tabulation is difficult to use when the field of investigation is vast and a number of characteristics are studied simultaneously. In such a case we can make use of machine tabulation.

Example 11:

In certain data, the following four main characteristics with the sub-characteristics are present :

Main Characteristics	***Sub-characteristics***
Locality	*Urban Rural*
Religion	*Hindu, Non-Hindu*
Sex	*Males, Females*
Age	*0–30, 30–60, 60 and over*

Prepare a suitable form of table.

Solution:

Table Showing the Sex-wise Distribution of the People of Different Religions, Localities and Age-groups

Age (In years)	**Hindus**								
	Urban			**Rural**			**Total**		
	M	**F**	**Total**	**M**	**F**	**Total**	**M**	**F**	**Total**
0–30	—	—	—	—	—	—	—	—	—
30–60	—	—	—	—	—	—	—	—	—
60 7 Over	—	—	—	—	—	—	—	—	—
Total	—	—	—	—	—	—	—	—	—

Age (In years)	**Non-Hindus**								
	Urban			**Rural**			**Total**		
	M	**F**	**Total**	**M**	**F**	**Total**	**M**	**F**	**Total**
0–30	—	—	—	—	—	—	—	—	—
30–60	—	—	—	—	—	—	—	—	—
60 7 Over	—	—	—	—	—	—	—	—	—
Total	—	—	—	—	—	—	—	—	—

Example 12:

The data given below relate to the heights and weights of 20 persons. You are required to form a two-way frequency table with class interval 62" to 64", 61" to 66", and so on and 115 to 125 lb., 125 to 135 lb., etc.

S. No.	*Weight*	*Height*	*S. No.*	*Weight*	*Height*
1	*170*	*70*	*11*	*163*	*70*
21	*135*	*65*	*12*	*139*	*67*
3	*136*	*65*	*13*	*122*	*63*
4	*137*	*64*	*14*	*134*	*68*
5	*148*	*69*	*15*	*140*	*67*
6	*124*	*63*	*16*	*132*	*69*
7	*117*	*65*	*17*	*120*	*66*
8	*128*	*70*	*18*	*148*	*68*
9	*143*	*71*	*19*	*129*	*67*
10	*129*	*62*	*20*	*152*	*67*

Using standard deviation and its coefficient, state whether there is a greater variation in height or weight.

Solution:

As per the requirements of the question, the population is to be divided into five classes according to the heights of the persons included in each group and six classes according to the weight. There will be thus 5 × 6 = 30 cells.

For tabulating the information in appropriate cells, first, the row to which the height measurement (say, X) should belong is determined. Afterwards on a consideration of the weight (say, Y), the column in which it should be included is determined. The tabulation is recorded by tally bars. Thus the two way table shall be prepared like this.

Two-way Frequency Table Showing Weight and Height of 20 Persons

Weight in lbs. (Y) / Height in inches (X)	**115–125**	**125–135**	**135–145**	**145–155**	**155–165**	**165–175**	**Total**
62–64	‖ (2)	| (1)					3
64–66	| (1)		||| (3)				4
66–68	| (1)	| (1)	‖ (2)	| (1)			5
68–70		‖ (2)		‖ (2)			4
70–72		| (1)	| (1)		| (1)	| (1)	4
Total	4	5	6	3	1	1	20

2

GRAPHIC PRESENTATION OF DATA

INTRODUCTION

Here we will discussed some method by tabulation and classification of data with the help of graph.

The most convincing and appealing ways in which statistical results may be presented is through diagrams and graphs. Evidence of this can be found in newspapers magazines, journals, advertisements, etc. There are numerous ways in which statistical data may be displayed pictorially such as different types of diagrams, graphs and maps. Very often the problem is that of selecting the best out of several methods that may be available. This is a difficult task and requires a great deal of artistic talent and imagination on the part of the individual or agency engaged in the preparation of diagrams and graphs. It is not practicable to discuss all the possible forms of charts here. An attempt is made in this chapter to illustrate some of the major types of diagrams, graphs and maps frequently used in presenting statistical data.

SIGNIFICANCE OF DIAGRAMS AND GRAPHS

Graphs are very useful due to following reasons:

(a) They facilitate comparison of data relating to different periods of time of different regions. Diagrams help one in making quick and accurate comparison of data. They bring out hidden facts and relationship and can stimulate as well as aid analytical thinking and investigation.

(b) They have a great memorising effect. The impressions created by diagrams last much longer than those created by the figures presented in a tabular form.

(c) They are attractive to the eye. Figures are dry but diagrams delight the eye. For this reason diagrams create greater interest than cold figures. Thus, while going though journals and newspapers the readers generally skip over the figures but most of them do look at the diagrams and graphs. Since diagrams have attraction value, they are very popular in exhibitions, fairs, conferences, board meetings and public functions.

(d) It is a fact that as the number and magnitude of figures increases they become more confusing and their analysis tends to be more strenuous. Pictorial presentation helps in proper understanding of the data as it gives an interesting form to it. The old saying A picture is worth 10,000 words is very true. The mind through the eye can more readily appreciate the significance of figure in the form of pictures than it can follow the figures themselves.

Comparison of Tabular and Diagrammatic Presentation

Tabulat and diagrammatic presentation have their own usefulness for particular purposes. Hence, the choice of the form of presentation must be made with due thought and care. We should kept following point in mind:

(a) Tables contain precise figures whereas diagrams give only an approximate idea. Exact values can be read from a table.

(b) Graphs and diagrams have a visual appeal and, therefore, prove to be more impressive to laymen.

(c) More information can be presented in one table than either in one graph or diagram.

(d) Tables usually require much close reading and are more difficult to interpret than diagrams.

Difference Between Diagrams and Graphs

(c) Diagrams are more attractive to the eye and as such are better suited for publicity and propaganda. They do not add anything to the meaning of the data.

(a) A diagram is generally constructed on plain paper. In other words, a graph represents mathematical relationship (though not necessarily functional) between two variables whereas a diagram does not.

(c) For representing frequency distributions and time series graphs are more appropriate than diagrams. In fact for presenting frequency distributions diagrams are rarely used.

GENERAL RULES FOR CONSTRUCTING DIAGRAMS

We use the following rules to construc the diagrams:

1. **Footnotes:** In order to clarify certain point about the diagram footnote may be given at the bottom of the diagram.
2. **Index:** An index illustrating different types of lines or different shades, colours should be given so that the reader can easily make out the meaning of the diagram.
3. **Neatness and Cleanliness:** Diagrams should be absolutely neat and clean.

4. **Simplicity:** Diagrams should be as simple as possible so that the reader can understand their meaning clearly and easily.
5. **Title:** Every diagram must be given a suitable title. The title should convey in as few words as possible the main idea that the diagrams intend to portray.
6. **Proportion between width and height:** We should maintain the proper proportion between heighest and width. If either the height and width is too short or too long in proportion, the diagram would give an ugly look. *"Graphic Presentation"* may be adopted for general use It is know as "Root-two" that is, a ration of 1 (short side) to 1.414 (long side). Modifications may no doubt, made to accommodate a diagram in the space available.
7. **Selection of Scale:** The scale showing the values should be in even numbers or in multiples of five or ten *e.g.,* 25, 50, 75, or 20, 40, 60. Odd values like 1. 3. 5, 7 should be avoided. Again no rigid rules can be laid down about the number of rulings on the amount scale, but ordinarily it should not exceed five. The scale should also specify the size of the unit and what it represents;

TYPES OF DIAGRAMS

Diagrams can be divided in the following ways:

1. Three-dimensional diagrams, *e.g.,* cubes, cylinders and spheres.
2. Two-dimensional diagrams, *e.g.,* rectangles, squares and circles.
3. One-dimensional diagrams, *e.g.,* bar diagrams.
4. Pictograms and cartograms.

One-dimensional or Bar Diagrams

A bar is a thick line whose width is shown merely for attention. They are called one-dimensional because it is only the length of the bar that matters and not the width. When the number of items is large lines may be drawn instead of bars to economise space. The special merits of bar diagrams are the following.

(i) They are readily understood even by those unaccustomed to reading charts or those who are not chart-minded.

(ii) They possess the outstanding advantage that they are the simplest and the easiest to make.

(iii) When a large number of items are to be compared they are the only form that can be used effectively.

To construct the bar kept the following points in mind.

(i) The gap between one bar and another should be uniform throughout.

(ii) The width of the bars should be uniform throughout the diagram.

(iii) While constructing the bar diagram, it is desirable to write the respective figures at the end of each bar so that the reader can know the precise value without looking at the scale. This is particularly so where the scale is too narrow for Ex. 1 on paper may represent 10 crore people. The two diagrams below would clarify the difference.

(iv) Bars may be either horizontal or vertical. The vertical bars should be preferred because they give a better book and also facilitate comparison.

The fixed assets of Tata Steals from 2001-2002 to 2003-04 arc given below:

year	:	1993-94	1994-95	1995-96
Fixed Assets (Rs. in lakhs)	:	415	481	607

Represent the above data by a bar diagram.

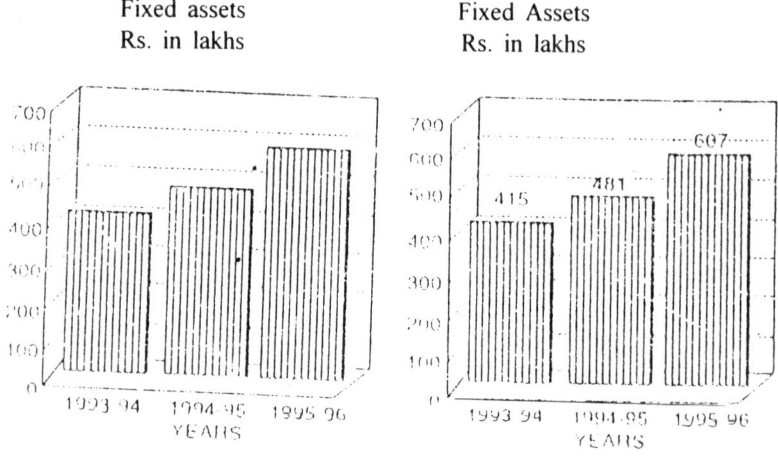

Types of Bar Diagrams

Bar diagrams can be classified as follows::

(a) Simple bar diagrams

(b) Sub-divided bar diagrams

(c) Multiple bar diagrams

(d) Percentage bar diagrams

(e) Deviation bars

(a) **Simple Bar Diagrams :** Simple bar diagram is used to represent only one variable. For Ex., the figures of sales, production, population etc.

for various years may be shown by means of a simple bar diagram. Since these are of the same width and only the length varies. Simple bar diagrams are very popular in practice. This can be either vertical or horizontal. In practice, vertical bars are more popular.

Example 1:

The growth production of Fish for the period 1991-92 to 1997-98 is given below:

Year	*Marine*	*Inland*	*Total*
1991-92	*5.34*	*2.18*	*7.52*
1992-93	*8.80*	*2.80*	*11.60*
1993-94	*10.86*	*6.70*	*17.56*
1994-95	*15.55*	*8.87*	*24.42*
1995-96	*16.98*	*11.03*	*28.01*
1996-97	*17.16*	*11.60*	*28.76*
1997-98	*12.47*	*8.42*	*20.89*

Represent the data by a suitable diagram.

Solution:

The above data can be represented by a sub-divided bar diagram drawn on a vertical base as follows :

Growth of production of fish 1991-92 to 1997-98

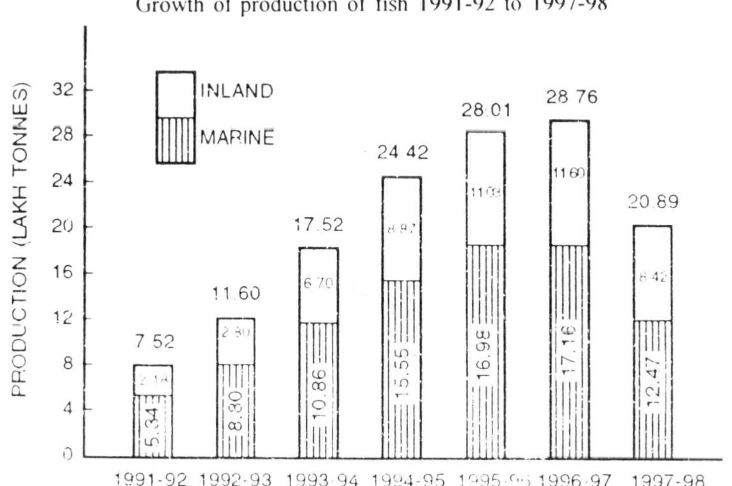

Example 2:

Following table gives the birth rate per thousand of different countries over a certain period:

Country	*Birth Rate*	*Country*	*Birth Rate*
India	*33*	*China*	*40*
Germany	*16*	*New Zealand*	*30*
U.K.	*20*	*Sweden*	*15*

Represent the above data by a suitable diagram

Solution:

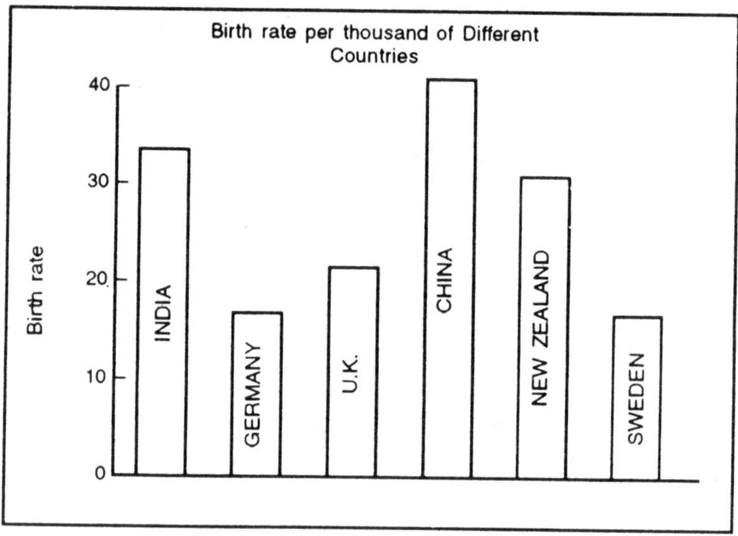

(b) **Sub-divided Bar Diagrams:** In a sub-divided bar diagram each bar representing the magnitude of a given phenomenon is further sub-divided in its various components. Each component occupies a part of the bar proportional to its share in the total.

This type of bar diagrams should not be used where the number of components is more than 10 or 12, for in that case, the diagram will be overloaded with information which cannot be easily compared and understood.

The component bar diagram can be used to represent either the absolute data or distribution ratios such as percentage distributions. It is, in fact, an excellent method for presenting a set of distribution ratios diagrammatically. The sub-divided bar diagrams can be constructed both on horizontal and vertical bases.

Example 3:

The regional rainfall indices during the year are given below:

Zone

Year	*West*	*North*	*East*	*South*	*Centre*
1996	*78.4*	*88.9*	*83.7*	*89.9*	*86.5*
1997	*75.6*	*62.5*	*103.6*	*75.5*	*77.4*
1998	*121.2*	*116.5*	*107.6*	*123.9*	*90.3*

Represent the data by a multiple bar diagram.

Solution:

The above data can be represented by following bar diagram.

Zone-wise Rainfall for 1996-1998

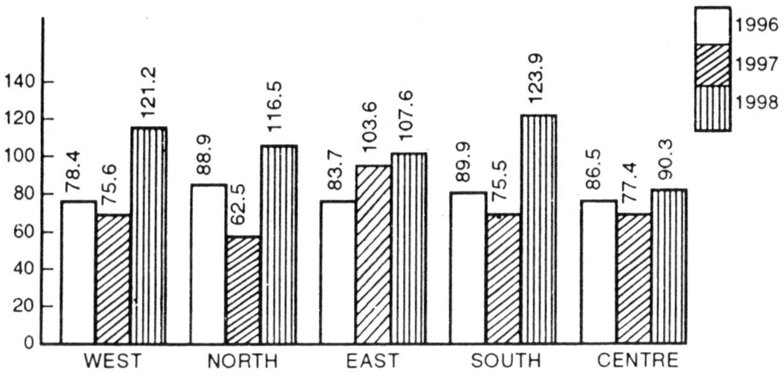

Example 4:

The following data relate to exports in 1992-93 (provisional) and export target for 1993-94

	1992-93	***Figure (Rs. crore)*** *1993-94*
Textiles	*1'5185*	*23680*
Gems and jewellery	*8839*	*11056*
Engineering goods	*6505*	*8000*
Agricultural products	*5514*	*7216*
Chemical products	*5419*	*6938*
Leather products	*3692*	*4768*
Ores and minerals	*2146*	*2781*
Marine products	*1743*	*2080*

Plantations	*1339*	*1600*
Electronics	*611*	*1200*
Petroleum products	*1379*	*861*
Raw cotton	*196*	*544*

Represent the data by horizontal bars.

Solution:

Exports Target for 1993-94: Government looking for 33% increase in exports in 1993-94.

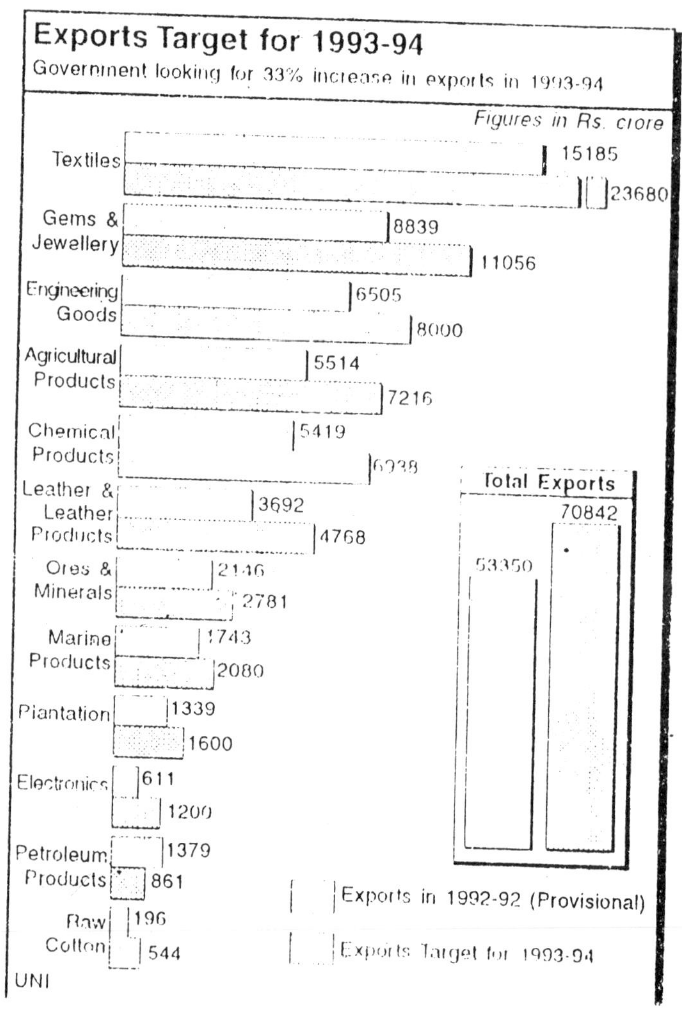

(c) **Multiple Bars:** In a multiple bar diagram two or more sets of interrelated data are represented. The technique of drawing such a diagram is the same as that of simple bar diagram. The only difference is that since more than one phenomenon is represented, different shades, colours, dots or crosses are used to distinguish between the bars. Wherever a comparison between two or more related variables is to the made, multiple bar diagram should be preferred.

Example 5:

Draw a multiple bar diagram from the following data:

Year	*Sales ('000 Rs.)*	*Gross Profit ('000 Rs.)*	*Net Profit ('000 Rs.)*
1992	*120*	*40*	*20*
1993	*135*	*45*	*30*
1994	*140*	*55*	*35*
1995	*150*	*60*	*40*

Solution:

The above data can be represented by following bar diagram.

Sales, Gross Profits & Net Profits (For the year 1992-95)

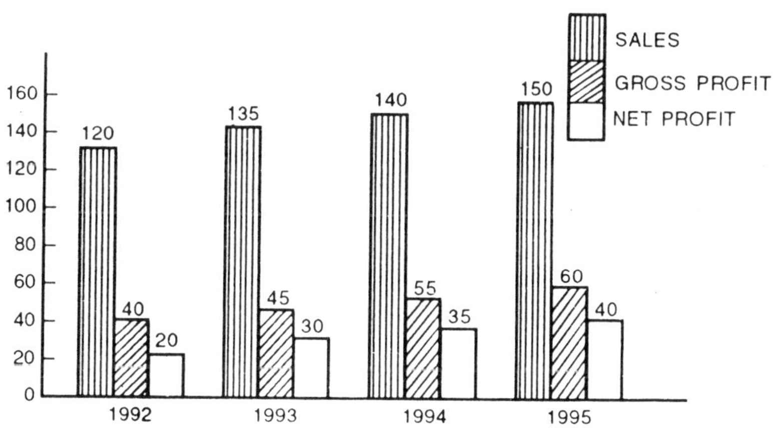

(d) **Percentage Bars:** Percentage bars are particularly useful in statistical work which requires the portrayal of relative changes in data When such diagrams are prepared, the length of the bars is kept equal to 100 and segments are cut in these bars to represent the components (percentages) of an aggregate.

Example 6:

Represent the following by sub-divided bar diagram drawn on the percentage basis:

Particulars	*1996*	*1997*	*1998*
1. Cost per chair			
(a) Wages	*9*	*15*	*21*
(b) Other	*6*	*10*	*14*
(c) Polishing	*3*	*5*	*7*
Total	*18*	*30*	*42*
2. Proceeds per chair	*20*	*30*	*40*
3. Profit (+) Loss(–)	*+2*	*–*	*–2*

Solution:

Take the sale price per chair as Rs. 100 and express the other figures in percentages. The percentages so obtained are given below:

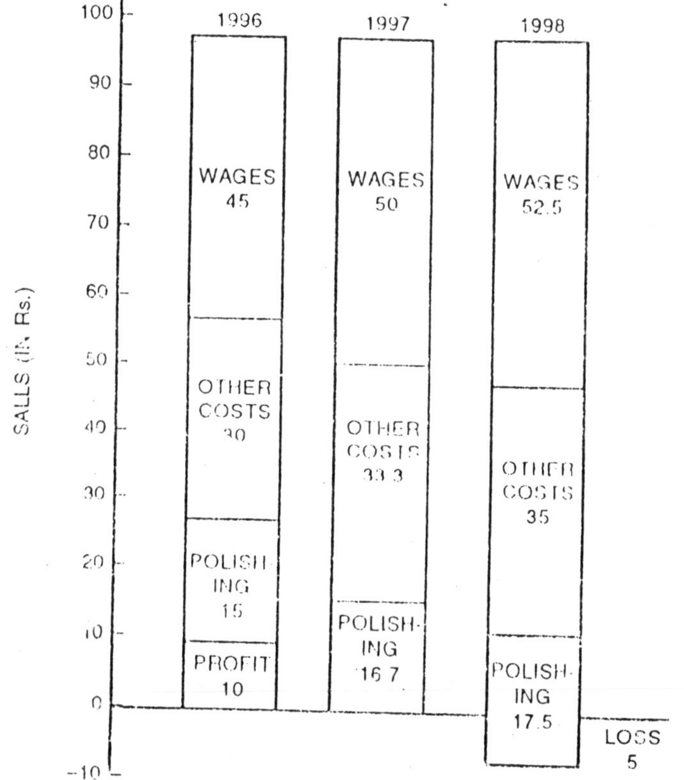

Particulars	*1996*	*1997*	*1998*
Wages	45.0	50.0	52.5
Other costs	30.0	33.3	35.0
Polishing	15.0	16.7	17.5
Total costs	90.0	100.0	105.0
Sale price	100.0	100.0	100.0
Profit or loss	+10.0	–	–5.0

(e) **Deviation Bars:** Deviation bars are popularly used for representing net quantities—excess or deficit, *i.e.,* net profit, net loss, net exports or imports, etc. Such bars can have both positive and negative values. Positive values are shown above the base line and negative values below it. The following illustration would explain this type of diagram:

Example 7:

The following data relate to the Additional Taxation during the year 1988-89 to 1997-98

Year	*Additional Taxation (Rs. in Crores)*
1988-89	*282*
1989-90	*271*
1990-91	*533*
1991-92	*716*
1992-93	*273*
1993-94	*431*
1994-95	*488*
1995-96	*514*
1996-97	*615*
1997-98	*1287*

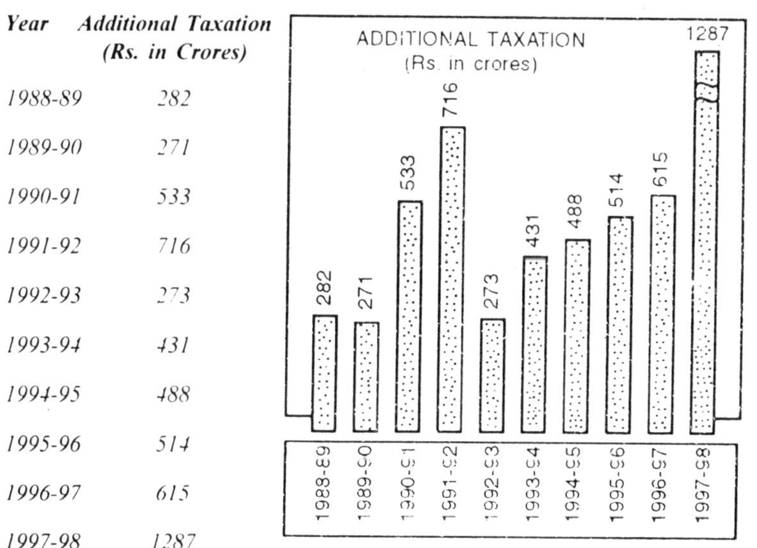

Represent the data by a suitable broken bar diagram.

Solution:

Since the gap between the minimum and maximum figure is large, the broken bar diagram shall be more appropriate.

TWO-DIMENSIONAL DIAGRAMS

As distinguished from one-dimensional diagrams in which only the length of the bar is taken into account, in two-dimensional diagrams the length as well as the width of the bars is considered. This the area of the bars represents the given data. Two-dimensional diagrams are also known as surface diagrams or area diagrams. The important types of such diagrams are: (a) Rectangles, (b) squares, and (c) Circles.

(a) **Rectangles:** When two sets of figures are to be represented by rectangles either of the two methods may be adopted. We may represent the figures as they are given or may convert them to percentage and then sub-divide the length into various components. The letter method is more popular than the former as it enables comparison to be made on a percentage basis. The following Ex.s would illustrate both these methods of constructing rectangular diagrams:

Example 1:

The following data relate to the monthly expenditure (in rupees) of two families A and B.

Items of Expenditure	*Expenditure (in Rs.)* *Family A*	*Family B*
Food	*1600*	*1200*
Clothing	*800*	*600*
Rent	*600*	*500*
Light and Fuel	*200*	*100*
Miscellaneous	*800*	*600*

Represent the above data by a suitable percentage diagram.

Solution:

Convert the given figures into percentages as follows:

Items of Expenditure	*Family A* *Rs.*	*%*	*cum Y*	*Family A* *Rs.*	*Y %*	*cum %*
Foot	1600	40	40	1200	40.00	40.00
Clothing	800	20	60	600	20.00	60.00
Rent	600	15	75	500	16.67	76.67
Light and Fuel	200	5	80	100	3.33	80.00
Miscellaneous	800	20	100	600	20.00	100.00
Total	4000	100		3000	100	

The area diagram is more difficult to read than to construct because of the problem of judging areas.

Percentage Rectangle Diagram

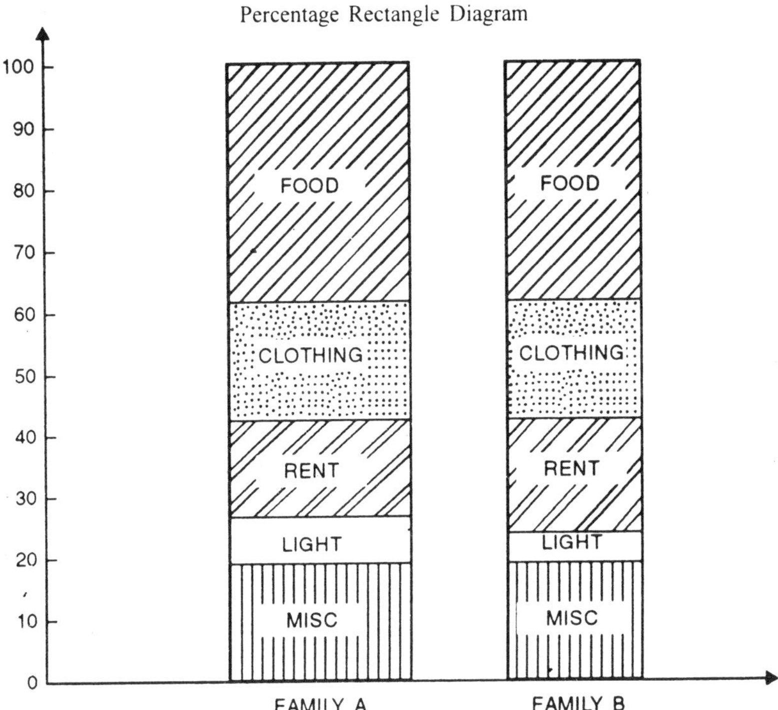

(b) **Squares:** The method of drawing a square diagram is very simple. One has to take the square-root of the values of various items that are to be shown in the diagrams and then select a suitable scale to draw the squares.

Example 2:

Represent the following data by a suitable diagram:

Total Public Sector Outlays

(Rs. '000 crores)

Five-year Plans	I	II	III	IV	V	VI	VII	VIII
Outlay	1.960	4.672	8.577	16.566	35.595	69.380	1.80.000	3.25.000

Solution:

Since three is a very big gap between the First Plan and Eighth Plan outlay, a square diagram may be quite suitable here. The size of one side of a square will be determined by the square root of the value to be represented and the size of the sides of various squares shall be proportional to the square roots of the various quantities to be presented.

Calculations for Drawing Square Diagram

PlanOutlay	Square-root	Side of the square in	
I	1.960	44.27	0.22
II	4.672	68.35	0.34
III	8.577	92.61	0.64
IV	16.566	128.71	0.64
V	35.595	188.67	0.94
VI	69.380	263.40	1.31
VII	1,80.000	424.26	2.12
VIII	3.25.000	570.08	2.85

Note: Each figure of the square-root has been divided by 200 and the side of the square obtained.

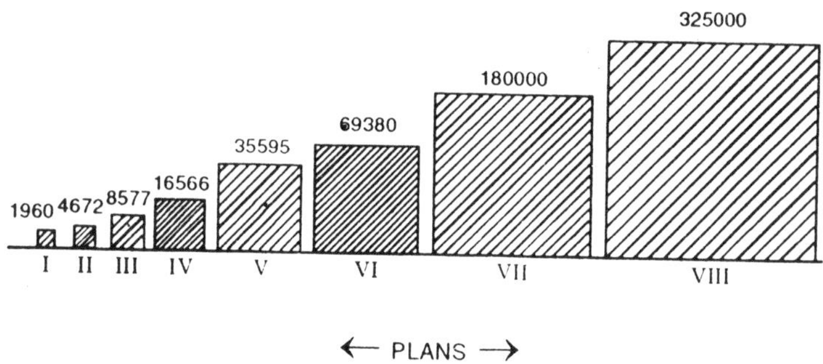

(c) **Circles:** Circles can be used in all those cases in which squares are used. However, in both these types of diagram it is difficult to judge the relative magnitudes with precision.

Example 3:

Represent the data of illustration 12 with the help of circles.

	Outlay	*Dividing* by $n - \left(\frac{22}{7}\right)$	*Square-root*
I	*1960.00*	*624.00*	*24.98*
II	*4672.00*	*1486.54*	*38.57*
III	*8577.00*	*2729.53*	*52.24*
IV	*16566.00*	*5271.80*	*72.63*
V	*35595.00*	*11325.68*	*106.41*
VI	*69380.00*	*22075.54*	*148.57*
VII	*1,80,000.00*	*57272.72*	*239.31*
VIII	*3,25,000.00*	*103409.4*	*321.57*

Note: *Area of a circle* $= \pi r^2$

$$r^2 = \frac{\text{Area}}{\pi}$$

$$r = \sqrt{\frac{\text{Area}}{\pi}} \text{ or } \sqrt{\frac{\text{Area}}{\pi}}$$

Solution:

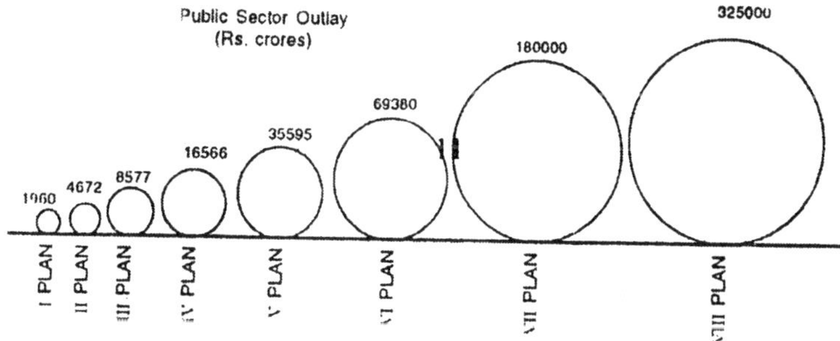

Circles are difficult to compare and as such are not very popular in statistical work. When it is necessary to use circles, they should be compared on an area basis rather than on a diameter basis, as the diameter basis is very misleading. Compared to rectangles, circles are more difficult to construct and interpret.

PIE DIAGRAM

While making comparisons, pie diagrams should be used on a percentage basis and not on an absolute basis, since a series of pie diagrams showing absolute figures would require that larger totals be represented by larger circles. Such presentation involves difficulties of two-dimensional comparisons. However, when pie diagrams are constructed on a percentage basis, percentages can be presented by circles equal in size. It may be noted that this problem does not arise in the use of a single pie diagram.

In laying out the sectors for pie chart, it is desirable to follow some largest component sector of pie diagram at 12 O'clock position on the circle. Usually the other component sectors are placed in clockwise succession in descending order of magnitude, except for catch-all components like "miscellaneous" and "all others" which are shown last, contrast with adjacent sectors.

In constructing a pie chart the first step is to prepare the data so that the various component values can be transposed into corresponding degrees on the circle. Suppose there are four components in a series representing the following values: (i) 60 per cent, (ii) 25 per cent, (iii) (360/100 = 3.6), the corresponding values of the four components in he illustration are (60) × (3.6) = 216; (25) × (3.6) = 90; (10) × (3.6) = 36; (5) × (3.6) = 18;

The second step is to draw a circle of appropriate size with a compass. The size of the radius depends upon the available space and other factors of presentation.

The third step is to measure points on the circle representing the size of each sector with the help of protractor. The ordinary protractor is based upon a scale in which the total circle is 360 degrees, but it is possible to purchase a protractor in which the entire circle is divided not into 360 but 100 equal parts so that the angle representing any desired percentage can be read directly.

In laying out the sectors for a pie chart it is desirable to follow some logical arrangement, pattern or sequence. For Ex., it is a common procedure to arrange the sectors according to size, with the largest at the top and others in sequence running clockwise. An essential feature of pie chart is the careful identification of each sector with some kind of explanatory or descriptive label. If there is sufficient room, the labels can be placed inside the sectors; otherwise the labels should be placed in contiguous positions outside the circle, usually with an arrow pointing towards the appropriate sector.

Example 1:

The following data relate to how the Rupee comes and goes.

Rupee comes	*Paise*	*Rupee goes*	*Paise*
Excise	*22*	*Central Plan*	*25*
Customs	*18*	*Interest*	*15*
Internal borrowing	*18*	*Defence*	*13*
None-tax revenue	*14*	*Share of taxes*	*14*
Deficit	*7*	*Other non-plan expenditure*	*12*
Other Capital receipts	*7*	*State & UT Plan assistance*	*12*
Corporation tax	*6*	*Subsidies*	*5*
Income-tax	*3*	*None-plan assistance*	*4*
External assistance	*3*		
Other taxes	*2*		

Solution:

Pie diagrams shall be appropriate here:

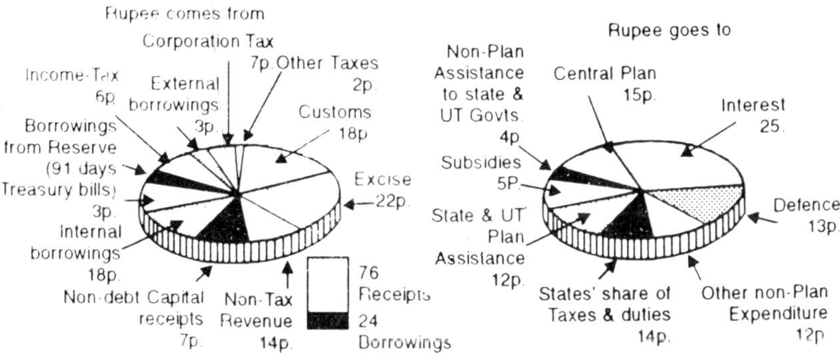

Limitations of Pie Diagrams

Pie diagrams at times are less effective than bar diagrams for accurate reading and interpretation, particularly when series are divided into a large number of components or the difference among the components is very small. It is generally inadvisable to attempt to portray a series of more than five or six categories by means of a pie chart.

Three-Dimensional Diagrams

Three-dimensional diagrams, also known as volume diagrams, consist of cubes, cylinders, spheres, etc. In such diagrams three things, namely, length, width and height, have to be taken into account. Such diagrams are used where the range of difference between the smallest and the largest value is very large.

Limitations of Three-Dimensional Diagrams

The three-dimensional diagrams have the same limiting features as the two-dimensional diagrams and to an even greater degree. Thus the side of a cube must be proportionate to the cube-root of the magnitude to be represented. It is very difficult for the eye to read precisely such diagrams and hence they are not recommended for statistical presentation.

Pictographs and Cartograms

Pictographs are very popularly used in presenting statistical data. They are not abstract presentations such as lines or bars but really depict the kind of data we are dealing with. Pictures are attractive and easy to comprehend and as such this method is particularly useful in presenting statistics to the layman. When pictographs are used data are represented through a pictorial symbol that is carefully selected.

While constructing a pictograph the following points should be kept in mind:

(a) The pictorial symbol should be self-explanatory. If we are telling a story about aeroplane, the symbol should clearly indicate an aeroplane. The following points should be kept in mind while selecting a pictorial symbol:

  (i) A symbol should suit the size of paper, *i.e.,* it should be neither too small nor too large.

  (ii) Last, but not the least, an artist should use the principles of a good design established by the fine and applied arts when drawing a pictorial symbol.

  (iii) A symbol must represent a general concept (like man, woman, child, bus) not an individual of the species (not Hitler. Akbar or Dr. Sharma's car).

  (iv) A symbol should be clear, concise and interesting.

  (v) A symbol must be clearly distinguishable from every other symbol.

(b) Changes in numbers are shown by more or fewer symbols, not by larger or smaller ones.

(c) Pictographs should be simple to understand and convey essential facts.

Example 2:

Draw a pie diagram for the following data of Sixth Five-Year Plan Public Sector outlays:

Agriculture and Rural Development	*12.9%*
Irrigation, etc.	*12.5%*
Energy	*27.2%*
Industry and Minerals	*15.4%*

Transport, Communication, etc. *15.9%*
Social Services and Others *16.1%*

Solution:

The angle at the centre is given by

$$\frac{\text{Percentage outlay}}{100} \times 360 = \text{Percentage outlay} \times 3.6$$

Computation for Pie Diagram

Sector	Percentage	Angle outlay
Agriculture and Rural Development	12.9	12.9 × 3.6 = 46°
Irrigation, etc.	12.5	12.5 × 3.6 = 45°
Energy	27.2	27.2 × 3.6 = 98°
Industry and Minerals	15.4	15.4 × 3.6 = 56°
Transport, Communication, etc.	15.9	15.9 × 3.6 = 57°
Social Services and others	16.1	16.1 × 3.6 = 58°
Total	100.0	360°

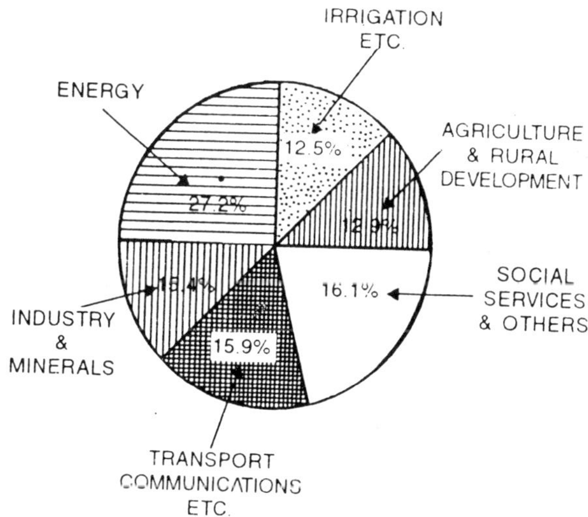

Now a circle shall be drawn suited to the size of the paper and divided into 6 parts according to degrees of angle at the centre. (The angles have been arranged in descending order).

Example 3:

The following table given the military balance between India and Pakistan:

	India	**Pakistan**
Men under arms	*11,3 (lakhs)*	*4.85*
Tanks	*2,400*	*1,600*
Artillery pieces	*2,500*	*1,400*
Armoured personnel carriers	*600*	*550*
Combat aircraft	*740*	*275*
Submarines	*8*	*11*
Aircraft carrier	*1*	*–*
Destroyers	*3*	*7*
Frigates	*21*	*16*
Naval combat aircraft	*36*	*3*

Represent the data by a pictogram

MILITARY BALANCE	*India*	*Pakistan*
Men Under Arms	11.3 Lakhs	1600 Lakhs
Tanks	2400	1600
Artillery Pieces	2500	4400
Armoured Personnel Carriers	600	550
Combat Aircraft	740	275
Submarines	8	11
Aircraft Carriers	1	–
Destroyers	3	7
Frigates	21	16
Naval Combat Aircraft	36	3

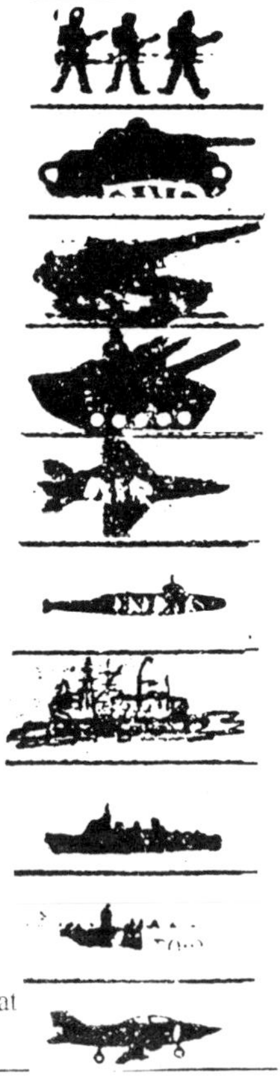

Merit : Compared with other types of diagrams, pictographs have a greater attraction value and, therefore, where the attention of masses is to be drawn such as in exhibitions, fairs, they are very popularly used. They stimulate interest in the information being represented.

Facts portrayed in pictorial form are generally remembered longer than facts presented in tables or in non-pictorial charts.

Limitations : They are difficult to construct. Besides, it is necessary to use one symbol to represent fixed number of units which may create difficulties. Thus if one symbol is representing 10 crore persons, the difficulties. Thus if one symbol is representing 10 crore persons, the question is how to represent a population of 31.85 crores. In such a case either the symbol should be proportionately smaller or the figure approximated to 30 crores. In either case, error is introduced.

Pictographs give only an overall picture, they do not give minute details. For greater accuracy we should write actual figure under, above or on one side of the symbol.

Cartograms : Cartograms or statistical maps are used to give quantitative information on a geographical basis. They are thus used to represent spatial distributions. The quantities on the map can be shown in many ways, such as through shades or colour, by dots, by placing pictograms in each geographical unit and by placing the appropriate numerical figure in each geographical unit. The cartogram given below shows the Telephone Network as on 1.6.99.

Statistical maps should be used only where geographic comparisons are of primary importance and where approximate measures will suffice. For more accurate representation of size, bar charts are preferable. To be sure, maps are sometimes combined, which are drawn in the appropriate areas.

Choice of a Suitable Diagram

The choice would primarily depend upon two factors, namely: (i) the nature of the data; and (ii) the type of people for whom the diagram is meant. On the nature of the data would depend whether to use one-dimensional, whether to adopt the simple bar or sub-divided bar, multiple bar or some other type. As already stated, a cubic diagram would be preferred to a bar if the magnitudes of the figures are very wide apart.

Different types of diagrams such as bars, rectangles, cubes, pictographs and pie charts have specialised uses. However, bar diagrams are most popular in practice. There are different types of bars and the appropriate type of bar chat can be divided on the following basis:

(a) Percentage composition bar charts are better suited where changes in the relative size of component figures are to be exhibited.

(b) Multiple bar charts should be used where changes in he absolute values of the component figures are to be emphasised and the over all total is of no importance.

(c) Simple bar charts should be used where changes in totals are required to be conveyed.

(d) Component bar charts are better suited where changes in totals as well as in the size of component figures (absolute ones) are required to be displayed.

However, multiple and component bar chart should be used only when there are not more than three or four components, as a large number of components make the bar charts too complex to enable worthwhile visual impression to be gained. When a large number of components have to be shown, a pie chart is more suitable.

A pie chart is particularly useful where it is desired to show the relative proportions of the figs. that go make up a ingle overall total.

However, pie charts cannot be used effectively where a series of figures is involved, as a number of different pie charts are not easy to compare. Nor should changes in the overall total be shown by changing the size of the 'pie'.

Occasionally, circles are used to represent size. But it is difficult to compare them and they should not be used when it is possible to use bars. This is because it is easier to compare the lengths of lines or bars than to compare areas or volumes.

Cubes should be used in those cases where the difference between the smallest and largest values to be represented is very large. In other cases, cubes should not be used because comparison is to difficult with the help of cubes.

Pictographs and cartograms are very elementary form of visual presentation. However, they are more informative and more effective than other forms for presenting data to the general public who, by and large, neither possess much ability to understand nor take interest in the less attractive forms of presentation.

GRAPHS

A large variety of graphs are used in practice. However, here we shall discuss only some important types of graphs which are more popular. Broadly, the various graphs can be divided under the following two heads:

(a) Graphs of time series, and

(b) Graphs of frequency distributions

Constructing charts and graphs is an art which can be acquired through practice. There are a number of simple rules, adoption of which leads to the effectiveness of the graphs. However, before discussing these rules the elementary procedure of constructing a graph is considered.

Technique of Constructing Graphs

For constructing graphs, we make use of graph paper. Two simple lines are first drawn which intersect each other at right angles. The lines are known as coordinate axes. The point of intersection is known as the point of origin or the zero' point. The horizontal line is called the axis of X or abscissa' and the vertical line the axis of Y or ordinate. The alternative appellations are Y-axis and X-axis respectively. The following are the two lines:

Y
+ 3
QUADRANT-II
x − ve
y + ve
+ 2
QUADRANT-I
x + ve
y + ve
+ 1
X'
− 3 − 2 − 1 0 + 1 +2 +3
X
− 1
QUADRANT-III
x − ve
y − ve
− 2
QUADRANT-IV
x + ve
y − ve
− 3
Y'

Coordinate system for plotting for Arithmatic Graphs.

In the above figure, O is the point of origin, XOX is the axis of X or the 'abscissa' and YOY" the axis of Y or the 'ordinate'. Both positive as well as negative values can be shown on the graph. Distances measured towards the right or upward from the origin are positive and those measured towards the left or downwards are negative.

The whole plotting area is divided into four quadrants as shown above. In quadrant *I*, both the values of X and Y are positive. In quadrant *II*, Y is positive, X is negative; in quadrant *III*, both X as well as Y are negative and in quadrant *IV*, X is positive whereas Y is negative. Since most business data are positive quadrant *I* is most frequently used.

It is conventional to take the independent variable on the horizontal scale and the dependent on the vertical scale. In case of time series, time is represented on the horizontal scale and the variable on the vertical scale. For each axis a convenient scale is chosen which represents the units of a variable. The choice is made in such a manner that the entire data are accommodated in the space available. The scale on X-axis and Y-axis need not be identical.

On the arithmetic line graph, the Y scale must begin at zero as origin. Thus, the X-axis always runs through this zero origin. The zero line is the base line and the curve is interpreted in terms of distance from this base line. In one special case, when we present graphically a series of changes from a norm of 100 per cent then the 100 per cent line is considered the base line.

Once the scale is chosen equal space would represent equal amounts in case of natural scale. However, in case of ratio scale, it is not so. No hard and fast rule can be laid now about the ratio of the scale on the abscissa and on the ordinate because much would depend upon the given data and size of the paper However, conventionally X-axis is taken 1½ times as long as Y-axis. But there is no rigidity about it.

After the choice of the scale is made the last step in constructing a graph is to plot the given data by taking the corresponding values of X and Y. The various points so obtained are then joined by straight lines.

Graphs of Time Series or Line Graphs

When we observe the values of a variable at different points of time, the series so formed is known as time series. The technique of graphic presentation is extremely helpful in analysing changes at different points of time. On the X-axis we generally take the time and on the Y-axis the value of the variable and join the various points by straight lines. The graph so formed is known as the line graph. Such graphs are most widely used in practice. They are the simplest to understand, easiest to make and most adaptable to many uses. They require the least technical skill and at the same time enable one to present more information of a complex nature in a perfectly understandable form than any other kind of chart. Many variables can be shown on the same graph and a comparison can be made.

Rules for Constructing the Line Graphs on Natural Scale

In constructing a graph of time series on natural scale the following points should be kept in mind:

(a) If on one graph more than one variable is shown, they should be distinguished by the use of think, thin, dotted lines, etc., or different colours be used. Every graph should be given a suitable title. The unit of time in which the variable under consideration is measured should be clearly stated in the title, *i.e.*, an indication should bc given as to whether the variable is measured as at a date.

(b) Join the various points with straight lines, not curves.

(c) Take the time on the X-axis (horizontal) and the variable on the Y-axis (vertical). The unit of time in which the variable under consideration is measured should be clearly stated in the title, *e.g.*, an indication should be given as to whether the years are calendar or financial or whether the variable is measured as at a date.

(d) Begin Y-axis with zero and select a suitable scale so that the entire data is accommodated in the space available. On the arithmetic scale equal magnitude must be represented by equal distances. This requirement is true for both the X-axis many represent 1,000 units whereas 1'' on X-axis may represent gap between 1998 and 1999. The scale should be so chosen that horizontal axis is longer than the vertical one. If the fluctuations in the variable are too small or if the lowest value of the variable is large, the false base should be used.

(e) Lettering on the graph, *i.e.*, indication of years, units, etc., should be done horizontally and not vertically so that in order to read what is written, it is not necessary to turn the graph from one side to another.

(f) Corresponding to the time factor plot the value of the variable and join the various points by straight lines (and not with curves). The points on the graph should not be indicated by circles or crosses rather dots should be used so that they disappear into line.

False Base Line

One of the fundamental rules while constructing graphs is that the scale on the X-axis should begin from zero. Where the lowest value to be plotted on the Y scale is relatively high and a detailed scale is required to bring bout the variations in all the data, starting the Y scale with zero introduces difficulties. For Ex., if we have a series of production figures over a number of years ranging from 15,000 units to 25,000 units, then starting with a zero origin would have one of two undesirable consequences: either (i) the necessarily large intervals (say 5,000 units) on the Y scale would make us lose sight of the extent of fluctuations in the curve; or (ii) a necessarily large graph to permit small intervals (say, 1,000 units) would entail a waste of a large part of he graph, in addition to poor visual communication.

The sol. is to break the Y scale. If the zero origin is shown then the scale is broken by drawing a horizontal wavy line (also called kinked or zigzag line) or a vertical wavy line between zero and the first unit on the Y would clarify the points.

The X-axis can also be broken in a similar manner as shown by the following diagram:

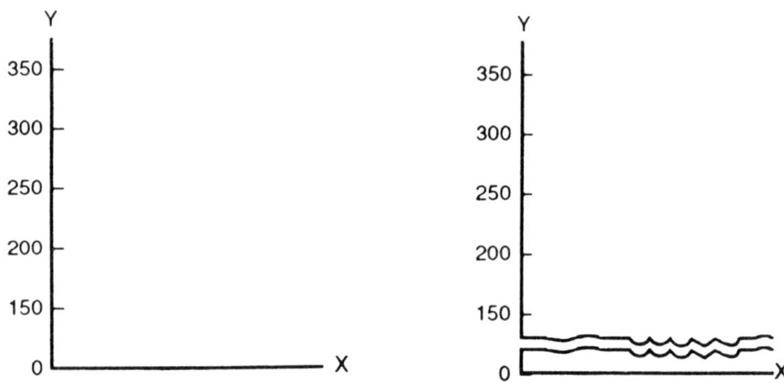

These lines are drawn to make the reader aware of the fact that false base has been used. Three important objects of false base line are:

(a) Variations in the data are clearly shown.

(b) A large part of the graph is not wasted or space is saved by using false base.

(c) The graph provides a better visual communication.

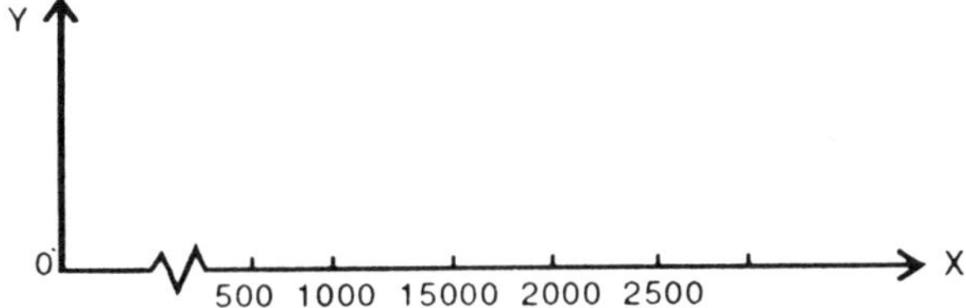

Graphs of One Variable

When only one variable is to be represented, on the X-axis measure time and on the Y-axis the value of the variable and plot the various points and join them by straight lines.

Example 1:

Represent the following data of per capita income graphically:

	Per Capita Income	*Year*	*Per Capita Income*
(At current price)			*(At current price)*
1987-88	*3285.4*	*1992-93*	*6261.7*
1988-89	*3842.1*	*1993-94*	*7195.7*
1989-90	*4346.5*	*1994-95*	*8402.6*
1990-91	*4983.0*	*1995-96*	*9578.4*
1991-92	*5602.9*	*1996-97*	*10771.2*

Source: Govt. of India : Economic Survey, 1997-98.

Solution:

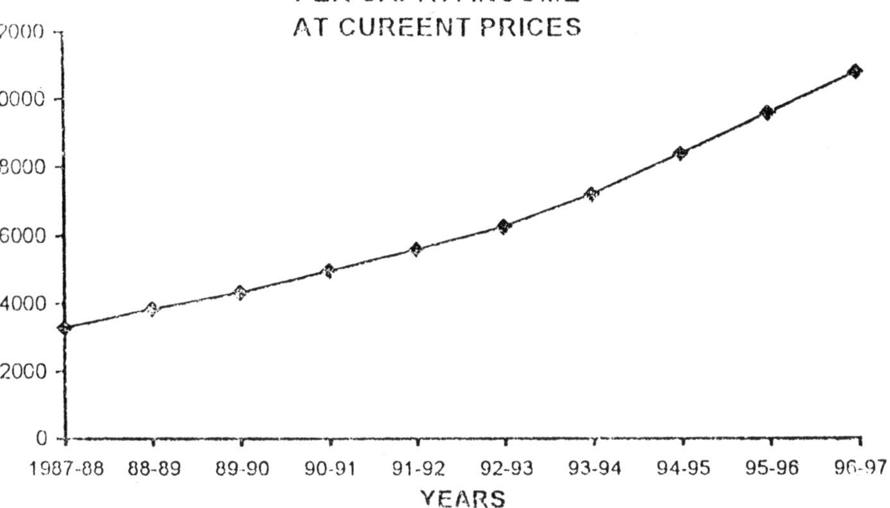

Example 2:

Represent the following data by a suitable graph

Per Capita National Produce (Rs.)

Year	*At Current Prices*	*At 1980-81 Prices*
1987-88	3284.4	1900.9
1988-89	3842.1	2059.0
1989-90	4346.5,	2157.1
1990-91	4983.0	2222.2
1991-92	5602.9	2175.1
1992-93	6261.7	2243.1
1993-94	7195.7	2337.2
1994-95	8402.6	2473.2
1995-96	9578.4	2608.2
1996-97	10771.2	2761.4

Graphs having Two Scales

If two variables are expressed in two different units, then we will have two scales–one on the left and the other on the right. To facilitate comparison, each scale is made proportional to the respective average of each. The average values of both the variables are kept in the middle of the graph, and then scales are determined.

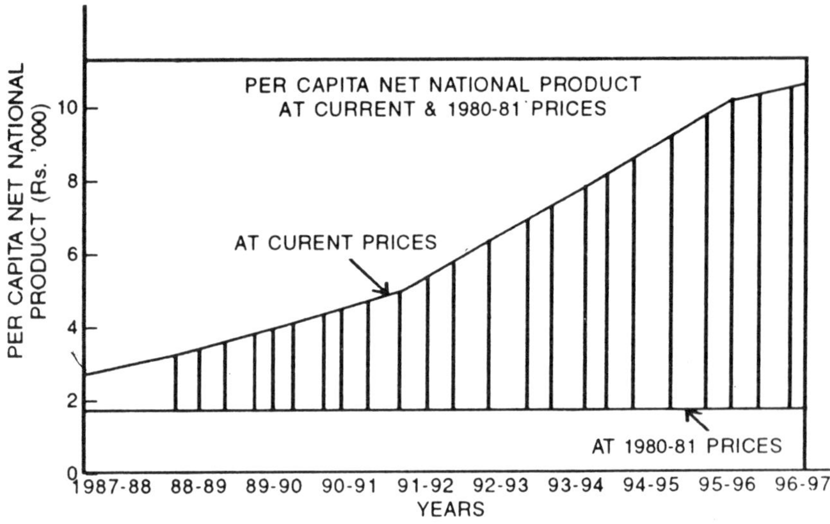

Example 3:

Represent the following data by a suitable graph

Year	*Index No. of Agricultural Production (Base 1981-82 = 100)*
1990-91	148.4
1991-92	145.5
1992-93	151.5
1993-94	157.3
1994-95	165.2
1995-96	160.7
1996-97	175.7
1997-98	164.9
1998-99	177.2

Graphs of Two or More Variables

If the unit of measurement is the same, we can represent two or more variables on the same graph. This facilitates comparison. However, when the number of variables is very large (say, exceeding five or six) and they are all shown on he same graph, the chart becomes quite confusing because

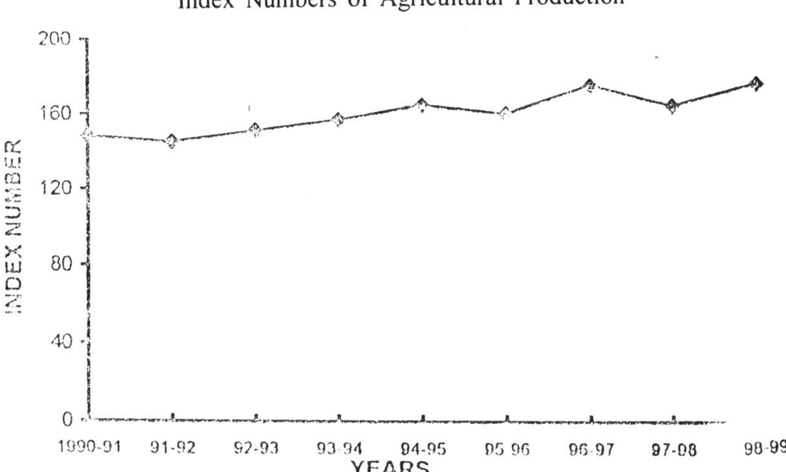

different lines may cut each other and make it difficult to understand the behaviour of the variables. Therefore, for the sake of clarity we should not represent more than 5 or 6 variables on the same graph. When two or more variables are shown on the same graph it is desirable to use thick, thin, broken, dotted lines, etc., to distinguish between the various variables.

Example 4:

Represent the following figures by a suitable diagram:

Year	*Principal imports (Rs. Crore)*	*Principal exports (Rs. Crore)*
1989-90	35328	27658
1990-91	43198	32553
1992-93	63375	53688
1993-94	73101	69751
1994-95	89971	82674
1995-96	122678	106353
1996-97	138919	118817

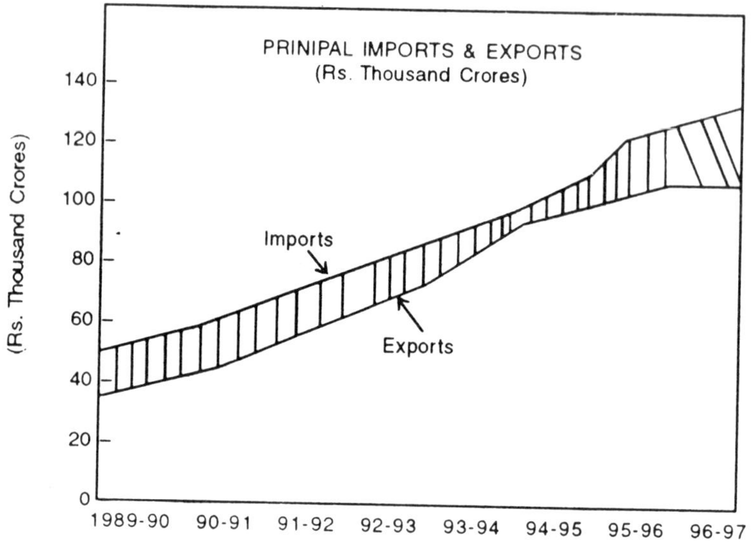

The above data can best be represented through a range chart, the following are the steps in constructing such a chart:

(a) Take time on the X-axis and the variable on the Y-axis.

(b) Draw two curves by plotting the given data–one curve representing the highest values and the other one the lowest values. In the given case, curve A represents lowest prices, whereas curve B highest prices. The gap between curve A and curve B represents the range of variation.

(c) For emphasising difference between the lowest and higher values the use of colour or some shade, etc., should be made.

Band Graph

A band graph is a type of line graph which shows the total for successive time periods broken up into sub-totals for each of the component parts of the total. In other words, the band graph shows how and in what proportion the individual items comprising the aggregate are distributed. The various component parts are plotted one over the other and the gaps between the successive lines are filled by different shades, colour, etc., so that the chart has the appearance of the series of bands. Such a chart is especially useful in dividing total costs into components costs, total sales into department or district or individual salesman's sales, total production by nature of commodity, states, plants or industries and other such relationships.

Band graph can also be used where the data are put to percentage form; the whole chart will depict 100 per cent and the bands the percentage each component bears to the whole.

Example 5:

The following data relate to sea-borne trade of wheat for one year:

Month	*Volume ('000 tonnes)*	*Value ('000 Rs.)*	*Month*	*Volume ('000 tomes)*	*Value ('000 Rs.)*
April	20	321	October	23	430
May	27	449	November	17	292
June	21	310	December	19	300
July	18	287	January	22	368
August	26	430	February	30	530
September	41	710	March	25	432

Represent the data by a suitable graph

Solution:

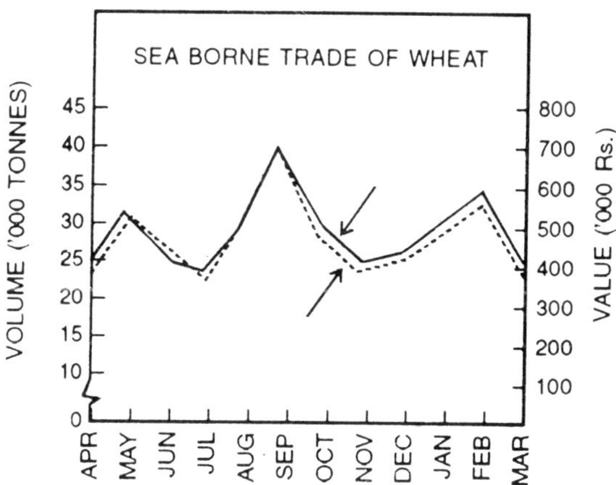

RANGE CHART

It is a very good method of showing the range of variation, *i.e.*, the minimum and maximum values of variable. For Ex., if we are interested in showing the minimum price of a commodity for different periods of time or the minimum and maximum temperatures, or the minimum and maximum price of shares of some company for different periods, the range chart would be very appropriate.

Example 1:

Represent the following data by a band graph:

Year	Rice	Wheat	Pulses	Other Cereals
1988-89	70.5	54.1	13.8	31.5
1989-90	73.6	49.8	12.8	34.8
1990-91	74.3	55.1	14.3	32.7
1991-92	73.7	55.1	12.0	26.3
1992-93	71.5	56.0	14.5	34.7

Solution:

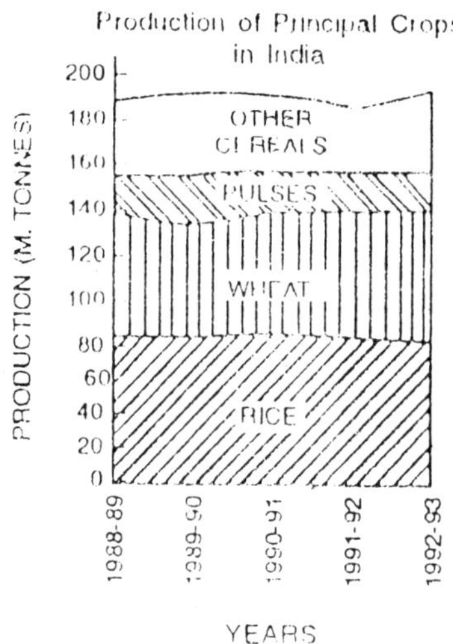

The steps in constructing the graph are given below:

(a) Take the years on the X-axis and the variables on the Y-axis.

(b) Plot the various points for different years for rice and join them by straight lines. This is represented by line A.

(c) Add the figures of rice for various years to the figures of wheat and plot the points and join them by straight lines. This is represented by line B. The difference between the two lines. *i.e.,* B and A, gives us the production of wheat.

(d) Add the figures of rice and wheat to pulses and plot the various points. This is represented by curve C. The difference between curve C and curve B represents production of pulses.

(e) Add the figures of rice, wheat and pulses to other cereals and draw a curve. This is represented by D. The difference between D and C gives the production figures for other cereals.

Semi-Logarithmic Line Graphs or Ratio Charts

The different types of graphs discussed so far have been drawn on natural or arithmetic scale. Sush graphs indicate the absolute changes in the values of a variable from one period to another. Thus if the profits of a firm rise from Rs. 1.000 to Rs. 2,000 and from Rs. 2,000 and from Rs. 2,000 to Rs. 3,000 in 3 years, on arithmetic scale the points would fall in a straight line thus indicating absolute change from one period to another, *i.e.,* Rs. 1,000 in each case. However, very often we are not interested in absolute amount of change in a variable rather our interest lies in ascertaining the rate at which the variable is increasing or decreasing. While studying sales, profits, production, etc., of a firm the absolute figures are not so important as the rate at which profits or sales, etc., are increasing or decreasing.

(a) In natural scale equal differences are measured by equal distances on the scale and thus, absolute movements are studied. In ration scale, however, the difference between scale measures equal proportional movement. This would be clear from the following example:

Natural scale	*Ratio scale*			
50	32	320	3200	32000
40	16	160	1600	16000
30	8	80	800	8000
20	4	40	400	4000
10	2	20	200	2000
0	1	10	100	1000

(b) The ration scale enables us to compare the rate of change of categories of different statistical units on the same chart. Many curves can be plotted on the same graph and their trends studied. For Ex., the trends of population, production of agricultural commodities, price, national income, employment, etc., can all be studied on a graph.

(c) In case of ratio scale, the Y-axis starts from one and not from zero whereas in the case of natural scale Y-axis starts from zero. The reason for the scale on semi-logarithmic paper starting at 1 and not zero is that the logarithm of 1 is 0, hence, value of 1 is placed at zero distance from the origin, *i.e.,* at the origin. There is no logarithm for zero, nor for negative numbers, hence such values cannot be plotted.

(d) It is clear from the above that the natural scale is based on the arithmetic progression whereas the ration scale is based on geometric progression.

(e) In case of ratio scale the meaning of the data is derived from the direction of lines whereas in case of natural scale the meaning is derived from position of lines.

(f) In case of variables having wide range of values the ratio scale graph is far more suitable than the other.

(g) Natural scale indicates absolute change, *i.e.*, equal distances on arithmetic paper represent equal amounts whereas ration scale indicates rate of change or the relative changes. In most of the problems of growth, absolute changes if shown on the graph are misleading. But the use of ration scale prevents one from drawing wrong conclusions.

Method of Constructing a Semi-Logarithmic Graph: A semi-logarithmic graph can be constructed in any of the following methods:

(a) By plotting the given values on a semi-logarithmic paper.

(b) By plotting the logarithms of the given values on natural scale.

When the first method is adopted the logarithms of the various values of variable are obtained by consulting the logarithmic tables. These logarithms are then plotted on the Y-axis of the natural scale and the various points are joined by straight lines to give us the required curve.

When the second method is adopted, we do not calculate the logarithms of the values of the variable, rather the actual values are plotted on the semi-logarithmic paper. This method is simple and convenient as compared to the first one because here one has not to calculate the logarithms of the values and hence there is considerable saving in time.

Example 2:

The following are the figures of sales of two Firms A and B for the years 1991 to 1998. Present the data graphically.

Year	*Sales Firm A (thousand units)*	*Sales Firm B (thousand) units)*	*Year*	*Sales Firm A (thousand)*	*Sales Firms B (thousand)*
1991	*200*	*2,000*	*1995*	*600*	*6,000*
1992	*300*	*3,000*	*1996*	*700*	*7,000*
1993	*400*	*4,000*	*1997*	*800*	*8,000*
1994	*500*	*5,000*	*1998*	*900*	*9,000*

Solution:

Let us plot the above date both on natural scale as well as on ration scale (*i.e.*, taking logs of various values) and compare the two graphs. The following is the graph of data on natural scale (graph A):

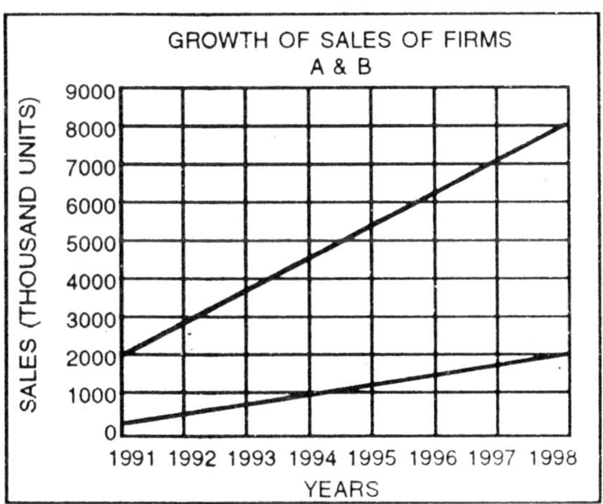

For presenting this data on ratio scale, taking logarithms of various values.

Year Firm A	*Sales logs 1000 units*	*Logs Sales*	*Firm B (1000 units)*	*Logs*
1991	200	2.3010	2.000	3.3010
1992	300	2.4771	3.000	3.4771
1993	400	2.6021	4.000	3.5021
1994	500	2.6990	5.000	3.6990
1995	600	2.7782	6.000	3.7782
1996	700	2.8451	7.000	3.8451
1997	800	2.8931	8.000	3.9031
1998	900	2.9542	9.000	3.9542

A comparison of both the above graphs ravels that when the data are plotted on natural scale, it shows a much higher rate of progress in case of firm b as compared to firm A. But when the data are potted on a ratio scale it indicates that the rate of growth is the same in both the firms. In fact, the sales of firm A are rising by 100 and that of firm B by one thousand progressive than firm A. However, the conclusion would be drawn only if

The above data is plotted below (graph B)

Growth Rate of Sales of firms A and B

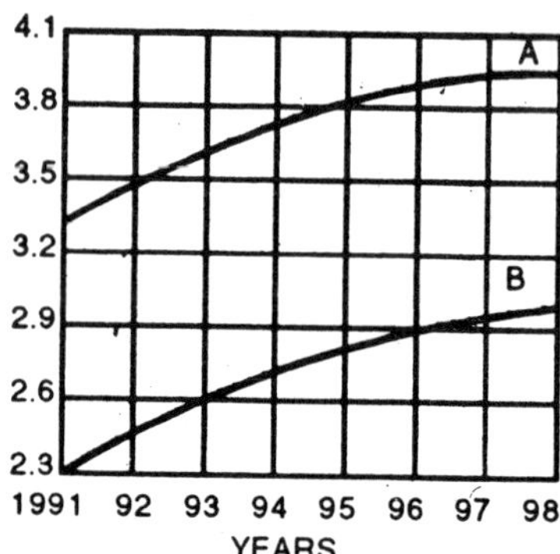

one would be clear that the rate of growth is the same in case of both firms A and B as is clearly shown by graph B.

Interpretation of Logarithmic Curves: The following are some of the important points that should be kept in mind while interpreting such curves:

(a) If a curve is falling but is nearly straight, it represents a decline at a nearly uniform rate.

(b) If a curve is rising but is nearly straight, it represents the growth at a nearly uniform rate.

(c) If a curve is a straight line, the rate of change is constant or uniform.

(d) If a curve is falling downwards, it represents a decreasing rate of change.

(e) If a curve is rising upwards, it would indicate an increasing rate of change.

(f) If one curve is steeper than another on the same ratio chart. The rate of change in the former is more rapid than that in the latter.

(g) If two curves on the same ration chart are found running parallel they represent equal percentage of change.

(h) If a curve is steeper in one portion than in another portion, the rate of change in the former is more rapid than that in the latter.

Uses of Ratio Charts: The ratio charts are useful as follows:

(a) The relative growth of fluctuations of two curves may be compared more accurately in ration charts then in arithmetic charts since parallel lines indicate the same per cent rates of change anywhere on the chart and steeper slopes indicates higher rates.

(b) By observing a company's production curve on a ratio chart the analyst can determine whether or not it is maintaining its past rate of growth. Furthermore, if historic factors of growth may be expected to persist, the analyst can project past trends, in order to forecast future volumes.

(c) Ratio scale is extremely useful in comparing series which differ widely in magnitude.

(d) Percentages or rations may be read directly from the vertical scale and applied toward further graphic analysis.

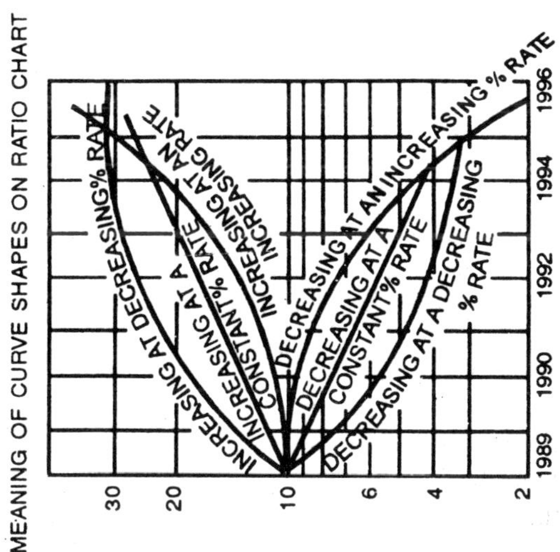

(e) A constant per cent rate of growth is represented by a straight line such as the sales increasing 10 percent a year appear on the ration chart as straight line. If the series curves away from the straight line it denotes a corresponding change in the rate of growth or the rate of decline as shown in the following chart on back page.

Limitations of Ratio Charts : The following are some of the limitations of such charts:

(a). The study of an aggregate into various component parts is not possible by using ratio scale.

(b) Zero or negative values cannot be shown on a ratio graph.

(c) They are difficult for the layman to understand and so should not be used for illustration which an arithmetic chart could show as well.

(d) Ratio scale cannot measure absolute changes.

(e) Their interpretation needs highly specialised knowledge in the absence of which one may draw entirely wrong conclusions. This factor alone restricts the scope of mass popularity of such a useful device.

GRAPHS OF FREQUENCY DISTRIBUTIONS

A frequency distribution can be presented graphically in any of the following methods:

(a) 'Ogives' or cumulative frequency curves. (b) Smoothed frequency curve, and (c) Frequency polygon, (d) Histogram;

Histogram

Histogram is the most popular and widely used in practice. A histogram is a set of vertical bars whose areas are proportional to the frequencies represented.

While constructing histogram the variable is always taken on the X-axis and the frequencies depending on it on the Y-axis. Each class is then represented by a distance on the scale that is proportional to its class-interval. The distance for each rectangle on the X-axis shall remain the same in case the class-intervals are uniform throughout. If they are different they vary. The Y-axis represents the frequencies of each class which constitute the height of its rectangle. In this manner we get a series of rectangles each having a class-interval distance as its width and the frequency distance as its height. The area of the histogram represents the total frequency as distributed throughout the classes.

The distinction lies in the fact that where a bar diagram is none dimensional, *i.e.,* only the length of the bar is material and not the width a histogram is two-dimensional, that is in a histogram both the length as well as the width are important.

We cannot construct a histogram for distribution with operend classes. Moreover, a histogram can be quite misleading if the distribution has unequal class-intervals and suitable adjustments in frequencies are not made.

The technique of contracting histogram is given below (i) for distributions have equal class-intervals, and (ii) for distributions having unequal class-intervals.

When class-intervals are equal, take frequency on the Y-axis, the variable on the X-axis and construct adjacent rectangles. In such a case the height of the rectangles will be proportional to the frequencies.

When class-intervals are unequal, a correction for unequal class intervals must be made. The correction consists of finding for each class the frequency density or the relative frequency density. The frequency density is the frequency for that class divided by the width of that class. A histogram or frequency density polygon constructed from these density values would have the same general appearance as the corresponding graphical display developed from equal class intervals.

For making the adjustment we take that class which has lowest class-interval and adjust the frequencies of other classes in the following manner. If one class-interval is twice as wide as the one having lowest class-interval we divide the height of its rectangle by two, if it is three times more we divide the height of its rectangle by three, etc., *i.e.*, the heights will be proportional to the ration of the frequencies of the width of the class.

Example 1:

A firm reported that its net worth in the year 1993-94 to 1997-98 was as follows:

Year	*1993-94*	*1994-95*	*1995-96*	*1996-97*	*1997-98*
Net worth	*100*	*112*	*120*	*130*	*147*

Plot the above data in the firm of a semi-logarithmic graph.

Solution:

To plot the data on a semi-logarithmic graph we will take the logs of the given values.

Year	*Net worth*	*Logs*
1993-94	100	2.0000
1994-95	112	2.0492
1995-96	133	2.0792
1996-97	133	2.1239
1997-98	147	2.1673

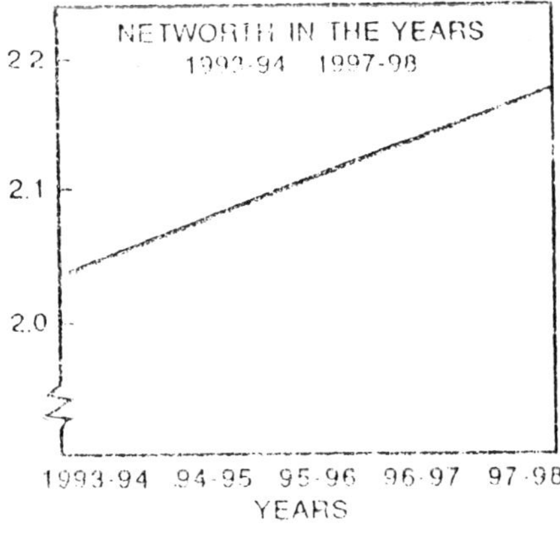

Example 2:

Draw the histogram for the following data:

Variable	*Frequency*	*Variable*	*Frequency*
100-110	*11*	*140-150*	*33*
110-120	*28*	*150-160*	*20*
120-130	*36*	*160-170*	*8*
130-140	*49*		

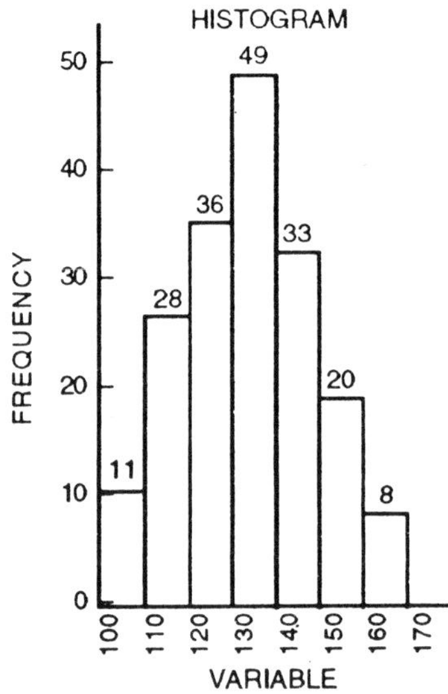

Frequency Polygon

It is particularly effective is comparing two or more frequency distributions. There are two ways in which a frequency polygon may be constructed:

(a) Another method of constructing frequency polygon is to take the mid-points of the various class-intervals and then plot the frequency corresponding to each point and to join all these points by straight by straight lines. The figures obtained would exactly be the same as obtained by method no. 1. The only difference is that here we have not to construct a histogram.

(b) We may draw a histogram of the given data and then join by straight lines the mid-points of the upper horizontal side of each rectangle with the adjacent ones. The figure so formed is called frequency polygon. It is an accepted practice to close the polygon at both ends of the distribution by extending them to the base line. When this is done two hypothetical classes at each end would have to be included-each with a frequency of zero. This extension is made with the object of making the area under polygon equal to the area under the corresponding histogram. The readers are advised to follow this practice.

(c) By constructing a frequency polygon the value of mode can be easily ascertained. If from the apex of the polygon a perpendicular is drawn on the X-axis, we get the value of mode. Moreover, frequency polygons facilitate comparison of two or more frequency distributions on the same graph.

Frequency polygon has following advantages over the histogram:

(a) The frequency polygon is simpler than its histogram counterpart.

(b) The frequency polygons of several distributions may be plotted on the same axis, thereby making certain comparisons possible, whereas histograms cannot be usually employed in the same way. To compare histograms we must have a separate graph for each distribution. Because of this limitation for purposes of making a graphic comparison of frequency distributions, frequency polygons are preferred.

(c) The polygon becomes increasingly smooth and curve-like as we increase the number of classes and the number of observations.

(d) It sketches an outline of the data pattern more clearly.

In the construction of frequency polygon the same difficulties are faced as with histograms, *i.e.,* they cannot be used for distributions having open-end classes and suitable adjustment, as in case of histogram, it is necessary when there are unequal class-intervals.

Example 1:

Represent the following data by means of a histogram:

Weekly Wages (in Rs.)	*No. of Workers*	*Weekly Wages (in Rs.)*	*No. of Workers*
10-15	*7*	*30-40*	*12*
15-20	*19*	*40-60*	*12*
20-25	*27*	*60-80*	*8*
25-30	*15*		

Solution:

Since the class-intervals are unequal, frequencies must be adjusted otherwise the histogram would give a misleading picture. The adjustment is done as follows. The lowest class interval is 5. The frequencies of the class 30-40 shall be divided by two since the class interval is double, that of 40-60 by 4, etc.

Construction of Histogram when only Mid-points are given : When only mid-points are given, ascertain the upper and lower limits of the various classes and then construct the histogram in the same manner.

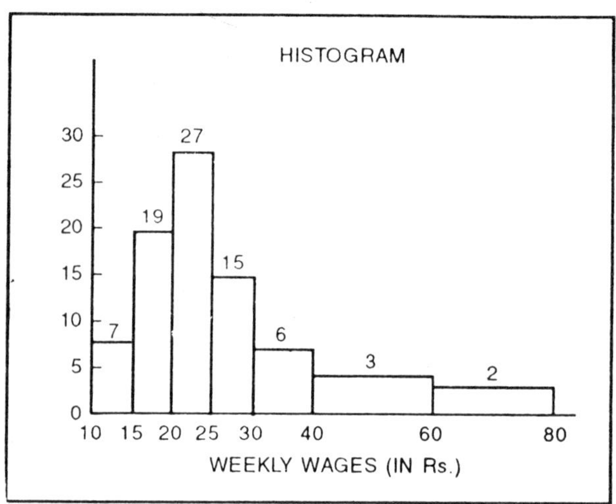

Example 2:

Represent the following data by a histogram:

Marks	*No. of Students*	*Marks*	*No. of Students*
0-10	*8*	*50-60*	*60*
10-20	*12*	*60-70*	*52*
20-30	*22*	*70-80*	*40*
30-40	*35*	*80-90*	*30*
40-50	*40*	*90-100*	*5*

Solution:

Since class-intervals are equal throughout no adjustment in frequencies is required.

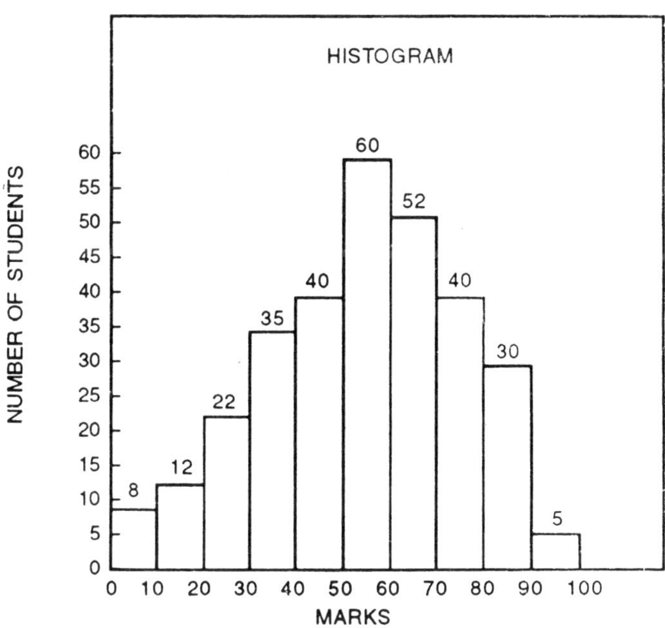

Example 3:

Draw a histogram and frequency polygon from the following data:

Marks	*No. of Students*	*Marks*	*No. of Students*
0-10	*4*	*50-60*	*14*
10-20	*6*	*60-70*	*8*
20-40	*14*	*70-90*	*16*
40-50	*16*	*90-100*	*5*

Solution:

Since class intervals are unequal, we shall have to adjust the frequencies. For Ex., the class 20-40 would be divided into two parts 20-30 and 30-40 with frequency of 7 each class.

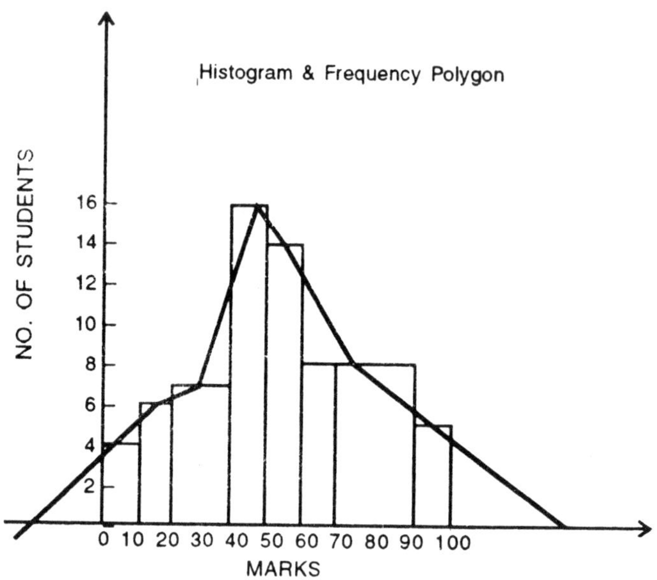

SMOOTHED FREQUENCY CURVE

The object of drawing a smoothed frequency curve is to eliminate as far as possible accidental variations that might be present in the data. While smoothing a frequency polygon the fact that it is really derived from the histogram should always be kept in mind. This would imply that the top of the curve would overtop the highest point of the polygon particularly when the magnitude of class-interval is large. The curve should look as regular as possible and sudden turns should be avoided. The extent of smoothing would, however, depend upon the nature of the data. If it is a natural phenomenon normally has symmetrical curves, but if the phenomenon is social or economic the curve is generally skewed and as such smoothing cannot be carried too far.

For drawing a smoothed frequency curve it is necessary to first draw the polygon and then smooth it out. As discussed earlier, the polygon can be constructed even without first constructing histogram by plotting the frequencies at the mid-points of class-intervals. This may save some time but the smoothing of the polygon cannot be done properly without a histogram. Hence it is desirable to proceed in a sequence, *i.e.,* first to draw a histogram then a polygon and lastly to smooth it to obtain the smoothed frequency curve. This curve should begin and end at the base line and as a general rule it may be extended to the mid-points of the class-intervals just outside the histogram. The area under the curve should represent the total number of frequencies in the entire distribution.

The following points should be kept in mind while smoothing a frequency curve:

(a) The total area under the curve should be equal to the area under the original histogram or polygon.
(b) Only continuous series should be smoothed.
(c) Only frequency distributions based on samples should be smoothed.

Example 1:

Draw less than and more than ogives from the data given below:

Profits (Rs. Lakhs)	*No. of Cos.*	*Profits (Rs. Lakhs)*	*No. of Cos.*
10-20	6	60-70	16
20-30	8	70-80	8
30-40	12	80-90	5
40-50	18	90-100	2
50-60	25		

Solution:

Less than ogive. In order to draw less then ogive we start with the upper limit of the classes as shows below:

Profit less than (Rs. lakhs):	20	30	40	50	60	70	80	90	100
No. of cos,	6	14	26	44	69	85	93	98	100

OGIVE BY 'LESS THAN' METHOD

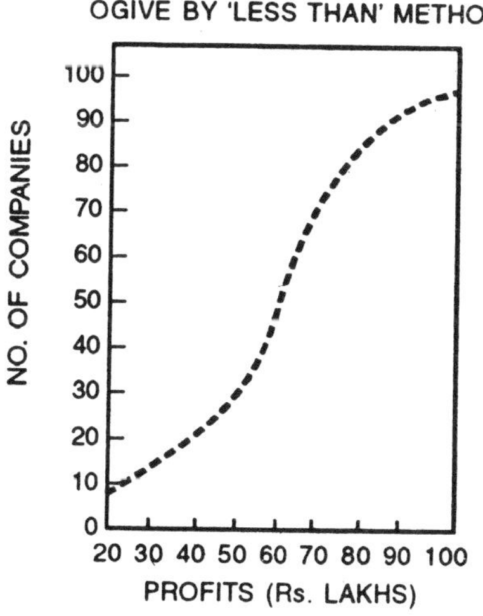

More than ogive: In order to draw more than ogive, we start with the lower limits of the various classes:

Profits (Rs. lakhs) More than	: 10	20	30	40	50	60	70	80	90
No. of companies	: 100	94	86	74	56	31	15	7	2

Example 2:

Represent the following frequency distribution by means of a Histogram and superimpose thereon the corresponding frequency polygon and frequency curve:

Salary (Rs.)	*No. of Employees*	*Salary (Rs.)*	*No. of Employees*
300-400	*20*	*700-800*	*115*
400-500	*30*	*800-900*	*100*
500-600	*60*	*900-1000*	*60*
600-700	*75*	*1000-1200*	*40*

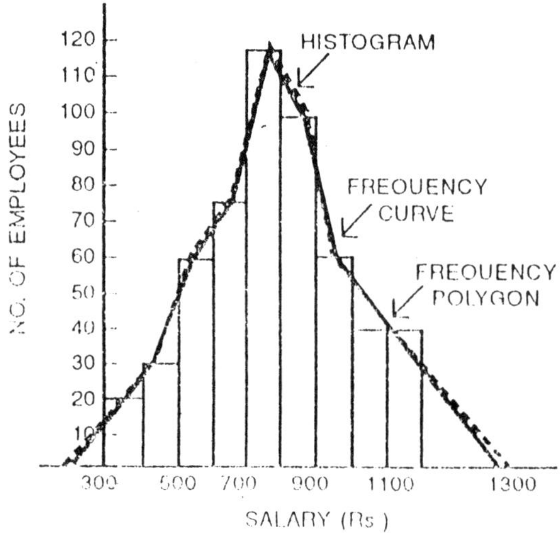

CUMULATIVE FREQUENCY CURVES OR OGIVES

At times we are interested in knowing 'how many workers of a factory earn less then Rs. 700 per month' or 'how many workers earn more than Rs. 1.000 per month' or percentage of students who have failed', etc. To answer these questions, it is necessary to add the frequents. When frequencies are added, they are called cumulative frequencies. These frequencies are then listed in a table called a cumulative frequency table. The curve obtained by plotting cumulative frequencies is called a cumulative frequency curve of an Ogive (pronounced Ojive).

There are two methods of constructing Ogive, namely:

1. The 'less than' method, and
2. The 'more than' method.

1. *'Less than' method.* In the 'less than' method we start with the upper limits of the classes and go on adding the frequencies. When these frequencies are plotted we get a rising curve.
2. *'More than' method.* In the 'more than' method we start with the lower limits of the classes and from the frequencies we subtract the frequency of each class. When these frequencies are plotted we get a declining curve.

The following frequency distribution is converted into a cumulative frequency distribution first by the 'less than' method and then by the 'more than' method:

Marks	*No. of Students*	*Marks*	*No. of Students*
10-20	4	40-50	20
20-30	6	50-60	18
30-40		60-70	2

Cumulative Frequency Distributions

Marks 'Less than'	*No. of Students*	*Marks 'More than'*	*No. of Students*
20	4	10	60
30	10	20	56
40	20	30	50
50	40	40	40
60	58	50	20
70	60	60	2
		70	0

From the above distribution one can read at once the number of students who have obtained marks less than a particular value or more than a particular value. Thus there are 20 students who have obtained marks less tan 40 and 50 students who have obtained marks more than 30.

Sometimes instead of writing 'Less than' and 'More than' we write 'or less' and 'or more'. The implication is different in the two cases. Thus marks less than 20 would exclude 20 whereas marks 20 or less' would include 20.

Similarly, marks more than 30 would exclude 30 whereas marks '30 or more' would include 30. One has to be very clear about the object in mind before these terms are used.

Utility of Ogives : The ogive is especially used for the following purposes:

(a) To compare two or more frequency distributions. Generally there is less overlapping when comparing several ogives on the same grid than when comparing several simple frequency curves in this manner.

(b) To determine as well as to portray the number or proportion of cases above or below a given value.

(c) Despite the great significance of ogives, it should be noted that they are not as simple to interpret as one may feel and hence the reader must be careful while using them.

(d) Ogives are also drawn for determining certain values graphically such as median, quartiles, deciles, etc.

Example 1:

The following table gives the wages of the workers in a certain factory:

Marks (Rs.)	*No. of workers*	*Daily wages*	*No. of workers*
20-25	*21*	*60-65*	*36*
25-30	*29*	*65-70*	*45*
30-35	*19*	*70-75*	*27*
35-40	*39*	*75-80*	*48*
40-45	*43*	*80-85*	*21*
45-50	*94*	*85-90*	*12*
50-55	*73*	*90-95*	*5*
55-60	*68*		

Draw a histogram and a frequency curve for the data given above. Find the number of workers whose wages lie between Rs. 57 and Rs. 77.

Solution:

The histogram of the above data is given

In order to find out the number of workers getting wages between Rs. 57 and Rs. 77 we will draw a less than ogive.

Daily wages (Rs.)	*No. of workers*	*Daily wages (Rs.)*	*No. of workers*
Less than 25	21	Less than 65	422
Less than 30	50	Less than 70	467
Less than 35	69	Less than 75	494
Less than 40	108	Less than 80	542
Less than 45	151	Less than 85	563
Less than 50	245	Less than 90	575
Less than 55	318	Less than 95	580
Less than 60	386		

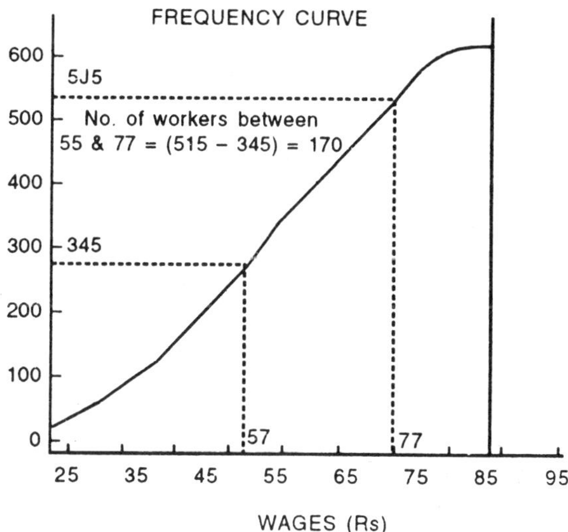

Limitations of Diagrams and Graphs

Julin has beautifully said, "Graphic statistics has a role to play of its own; it is not the servant of numerical statistics, but it cannot pretend, on the other hand, to precede or displace the latter"

The main limitations of diagrams and graphs are:

(a) They two-dimensional diagrams and the three-dimensional diagrams cannot be accurately appraised visually and, therefore, as far as possible their use should be avoided.

(b) They can approximately represent only limited amount of information.

(c) They can present only approximate values.

(d) They can be easily misinterpreted and, therefore, can be used for grinding one's axe during advertisement, propaganda and electioneering. As such diagrams should never be accepted without a close inspection of the bonafides because things are very often not what they appear to be.

(e) They are intended mostly to explain quantitative facts to the general public. From the point of view of the statistician, they are not of much help in analysing data.

Example 2:

Draw a percentage curve for the following distribution of marks obtained by 700 students in an examination:

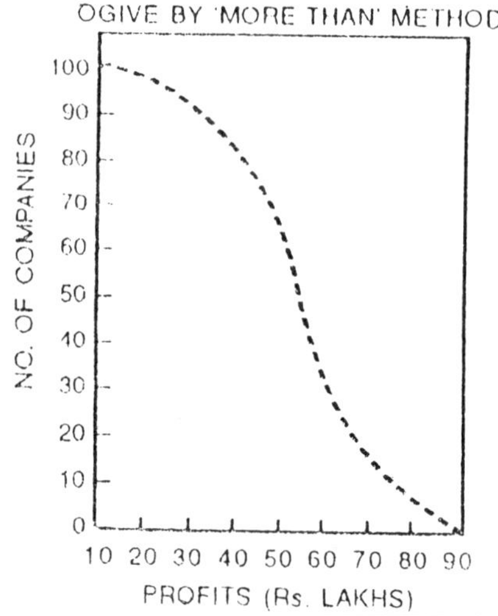

Marks	*No. of students*	*Marks*	*No. of students*
0-9	*9*	*50-59*	*102*
10-19	*42*	*60-69*	*71*
20-29	*61*	*70-79*	*23*
30-39	*140*	*70-79*	*2*
40-49	*250*	*80-89*	

Find form the graph (i) the marks at the 20th percentile, and (ii) the percentile equivalent to a mark of 65.

Solution:

A percentile curve is a cumulative curve drawn on a percentage basis. Hence for drawings such a curve three steps are required:

(a) Take these percentages on the Y-axis and the variable on the X-axis and plot the various points and join them by straight lines. The curve so drawn is known as the percentile curve.

(b) Convert these cumulative frequencies in percentage of the total.

(c) Find the cumulative frequencies of the given data by the 'less than' method.

Marks less than	*Frequency*	*Cumulative Frequency*	*Percentages*
9.5	9	9	1.3
19.5	42	51	7.3
29.5	61	112	16.0
39.5	140	252	36.0
49.5	250	502	71.7
59.5	102	604	86.3
69.5	71	675	96.4
79.5	23	698	99.7
89.5	3	700	100.0

It is clear from the graph that at 20th percentile marks are 31.5 and corresponding to 65 marks, the percentile is 90.

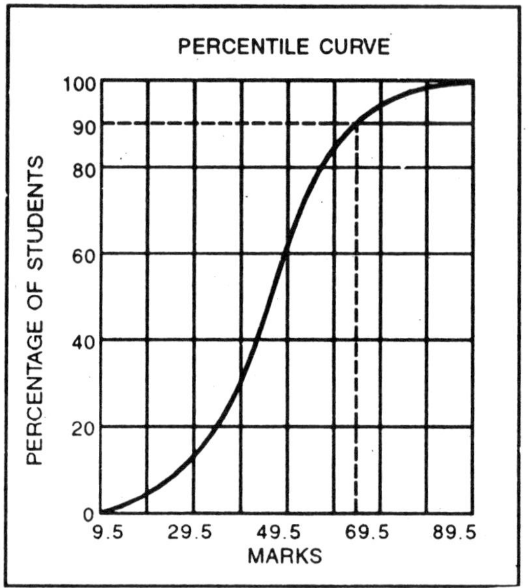

EXERCISES

1. The following table gives the country of origin of feature films exhibited in India:

Country :	India	U.S.A.	U.K.	Other Countries
No. of Films :	144	81	64	16

Represent them by Square or Circle diagram.

2. (a) What are 'ogive' curves? How are they used in reclassification of data?

   (b) "Graphs and diagrams are more effective than any other method of presenting data." Why?

3. (a) Discuss the merits and Imitations of representing statistical data through graphs and diagrams. How do you represent data by means of a pie diagram?

   (b) Point out the role of diagrammatic presentation of data.

   Explain briefly the different types of bar diagrams known to you.

4. (a) "Charts and graphs are more effective in attracting attention than any of the other methods of presenting data." Do you agree? Give reasons in support of your answer.

(b) What are the different methods of a graphical presentation of data? Explain them.

5. Represent the following data by a suitable diagram:

Item of expenditure	*Family A*	*Family B*	*Family C*
Food	50	45	60
Clothing	20	25	20
Rent	10	10	10
Education	5	10	5
Miscellaneous	15	10	5
Total	100	100	100

6. Represent the following data by a pie diagram:

Items	*Expenditure in Rupees*
Food	84
Clothing	27
Recreation	10
Education	15
Rent	23
Miscellaneous	21

7. Draw a histogram and a frequency polygon for the following data:

Class :	0-10	10-20	20-30	30-40	40-50	50-60	60-70
Frequency:	6	8	10	15	13	8	5

8. Represent the data given below with a suitable diagram.

Items	*Value in Rs. in 1997-98*	*Value in Rs. in 1998-99*
Raw material	150	200
Labour	100	200
Power	75	150
Advertisement	25	100
Other charges	50	200
Total cost of manufacturing	400	900

9. The annual profits in lakhs of rupees of 100 companies are distributed as follows: Profits per Co.

(Rs. lakhs) :	0-50	50-100	100-150	150-200	200-250	250-300
No. of cos :	12	18	27	20	17	6

Draw a histogram and frequency polygon.

10. The following table shows the Sixth Five-Year Plan public sector outlays by heads of development:

Heads of Development	***(Rs. Crores)*** ***Centre***	***States***
Agriculture	4,765	7,039
Irrigation and Flood Control	6,635	11,395
Energy	9,995	8,293
Industry and Minerals	12,770	2,985
Transport and Communication	12,200	5,120
Social Services	8,216	1,420
Total	54,581	36,252

11. The following table shows the results of B. Com. students of a college for the last three years. Present the data by a suitable diagram.

Years	***Ist Class***	***2nd Class***	***Pass***	***Failed***
1996	20	30	40	10
1997	30	50	20	20
1998	25	60	40	25

12. Draw histogram for the following data and superimpose on it the frequency curve:

Variable:	5-9	10-14	15-19	20-24	25-29	30-34	35-39
Frequency:	8	15	18	30	16	12	6

13. The following data relate to the expenditure of a family. Represent this data by angular (pie) diagram:

Items of expenditure	***Rupees (per month)***
Food	160
Clothing	110
Rent	80
Recreation	60
Miscellaneous	40

14. Draw a (rough) Pie chart to represent the following data relating to the production cost of manufacture:

Cost of Materials	:	Rs. 38,400
Cost of Labour	:	Rs. 30.720

Direct Expenses : Rs. 11.520

Overhead Expenses : Rs. 15.360

15. (a) What are different methods of graphical representation of data? Explain them.

    (b) Distinguish Histogram and Histogram clearly with illustrations.

    (c) What is histogram? How could you represent a grouped frequency distribution by means of a histogram when the class widths are (i) equal, and (ii) not equal.

16. (a) Explain the various methods that are used for graphical representation of frequency distribution.

    (b) What are the advantages of presenting data through diagrams and graphs?

    (c) Distinguish between diagrammatic and graphic representation of data.

17. (a) Illustrate graphically the distinction between a frequency polygon, a histogram and an ogive curve. Comment on their uses.

    (b) Prepare a 'more than ogive' curve with imaginary figures.

    (c) Explain the various diagrams which are used in statistics to show the salient features of data.

18. Represent the following data by a suitable diagram:

Item of expenditure	*Family A*	*Family B*	*Family C*
Food	*20*	*60*	*120*
Clothing	*4*	*15*	*70*
House Rent	*3*	*20*	*80*
Education	*2*	*6*	*25*
Books	*2*	*10*	*70*
Misc	*2*	*8*	*50*

19. Represent the following data by a suitable diagram:

    GNP–Industrial origin–percentages

Countries	*Agriculture*	*Industry*	*Services*	*Others*
England	3	40	44	13
America	3	35	61	1
Japan	6	48	43	3
India	45	19	28	8

20. Draw a histogram of the frequency distribution given below:

Variable	:	10-24	15-19	20-29	30-49
Frequency	:	5	10	30	20

Represent the data by some suitable diagram and write a report on the data bringing out the salient features.

21. (a) Indicate the method of constructing histogram, frequency polygon and ogive.

(b) Define 'Frequency polygon' and 'Frequency Curve'. Explain clearly the relation that exists between them.

22. (a) Distinguish between natural scale and ratio scale for drawing graphs. In which cases should the latter be used?

(b) Point out the significance of diagrams and graphs in the presentation of data.

23. (a) What is a histogram? How do you construct it?

(b) Discuss the utility of the graphic method of presenting statistical data. What points should be taken into account in the construction of graphs?

(c) Prepare 'more than' ogive curve with imaginary figures.

24. (a) What type of statistical data is best represented by a pie diagram? Illustrate your answer with Exs.

(b) "Diagrams do not add anything to the meaning of statistics but when drawn and studied intelligently, they bring to view the salient characteristics of graphs and series". Discuss the statement describing briefly the various types of diagrams.

25. (a) What is false base line? Under what conditions would its use be desirable?

(b) What conditions must generally be borne in mind while presenting statistical data?

26. (a) Explain with the help of sketches the construction of the following:

(i) Bar diagram,

(ii) Histogram,

(iii) Frequency polygon,

(iv) Circular diagram.

(b) Explain what is meant by a semilogarithmic diagram. Discuss the advantages of the natural scale diagram.

(c) What do you mean by a cumulative frequency distribution? Point its special advantages and uses.

27. Represent the following data relating to the monthly expenditure of two families A and B by means of rectangular diagram on a percentage basis:

Expenditure	*Family A income*	*Family B income*
Food	Rs. 2000	Rs. 2000
Clothing	600	480
Education	100	180
Fuel	140	60
House Rent	600	192
Miscellaneous	100	48

28. Represent the following data by an appropriate diagram:

	(In Crores of Rs.)	
Year	*Imports*	*Exports*
1997	1,600	2,000
1998	2,500	2,200
1999	2,800	2,400

29. Draw histogram, frequency polygon and frequency curve for the following distribution.

Experience (in months)	*No. of social workers*	*Experience (in months)*	*No. of social workers*
5-10	5	25-30	5
10-15	6	30-35	4
15-20	15	35-40	2
20-25	10	40-45	2

30. Which of the following statements are True of False?

(i) The area of a rectangle is equal to the product of its length and width. T/F

(ii) Cubes are two-dimensional diagrams. T/F

(iii) A frequency polygon has more than four sides. T/F

(iv) A pie diagram is a circle broken down into component sectors. T/F

(v) Data classified geographically or qualitatively cannot be presented on a line graph. T/F

(vi) A cumulative frequency distribution enables us to see how many observations lie above or below certain values. T/F

(vii) Squares are one dimensional diagrams. T/F

(viii) Bar diagram represents a frequency distribution. T/F

Ans. (i) T. (ii) F. (iii) T. (iv) T. (v) T. (vi) T. (vii) F. (viii) F.

31. Tick (✓) the correct answer:

(a) Diagram and graphs are tools of (i) collection of data. (ii) analysis. (iii) presentation. (iv) summarisation. (v) None of these.

(b) In a two-dimensional diagram (i) only height is considered. (ii) only width is considered, (iii) height, width and thickness are considered, (iv) both height and width are considered.

(c) Which of the following should be avoided as methods of presenting data: (i) spheres and cubs. (ii) bars, (iii) pie diagrams, (iv) petrographers. (v) rectangles? **Ans.:** (a) (ii). (b) (iv), (c) (i).

32. Draw a percentage subdivided rectangular diagram to represent the following data:

	Expenditure (in Rs.)	
	Family A (Rs.)	*Family B (Rs.)*
1. Food	2000	3000
2. Clothing	480	750
3. Education	320	400
4. House Rent	400	750
5. Travelling	600	900
6. Misc	200	200
Total Expenditure	4000	6000
Total Income	5000	8000
Total Savings	1000	2000

33. (a) Discuss various types of two dimensional diagrams. Explain with Ex., how these are prepared?

(b) Distinguish between 'Natural scale and 'Ratio Scale'. What are the methods of constructing graphs on ratio scale? How do you interpret graphs on ration scale? In which cases should ration scale the used?

34. (a) Discuss the methods of presentation of data through graphs and diagrams.

    (b) Discuss the various types of graphical presentation of data.

35. Proportions of males and females in India are given below according to occupation. Draw a suitable diagram:

Occupation	*Males*	*Females*
Manufacturing	47%	53%
Agricultural labour	55%	45%
Household industries	61%	39%
Miscellaneous	88%	12%

36. Present the following data of results of the II Yr. B. Com. Statistics examination of Bangalore University held in November, 1999 and November, 2000 by means of multiple bars:

Years	*I Class*	*II Class*	*III Class*	*Final*
November, 1998	100	300	500	300
November, 1999	120	400	600	280
November, 2000	100	500	700	300

37. Represent the following data with a suitable diagram:

Index Numbers of wholesale prices

Years	*Cereals*	*Pulses*	*Fibres*	*Oilseeds*
1992	443	424	432	499
1993	465	438	446	593
1994	471	449	476	665
1995	483	506	622	679
1996	450	483	454	483
1997	451	494	420	573

38. Draw a histogram of the following distribution:

Life of electric *Imps (in hrs.)*	1,010	1,030	1,050	1,070	1,090
Firm A:	13	130	482	360	18
Firm B:	287	106	26	230	352

39. Fill in the blanks:

    (i) "A picture is worth ...... words".

    (ii) For constructing a graphs, we generally make use of..... whereas a diagram is generally constructed on a......

(iii) Bar diagrams are ...... dimensional diagrams.

(iv) Cartograms are used to give quantitative information on a....

(v) Graphs of time series are called...... whereas graphs of frequency distribution are called.......

(vi) Natural scale indicates......changes whereas ratio scale indicates.......changes.

(vii) Point out which is correct:

Diagrams are for.......

(a) the use of experts.

(b) better mental appeal.

(c) None of these.

Ans.: (i) 10,000, (ii) graph paper, plain paper, (iii) one. (iv) geographical basis, (v) historigrams histograms, (vi) absolute, relative, (vii) better mental appeal.

40. (a) State the difference between Natural sale and Ration scale.

(b) What purpose is served by a semi-logarithmic graph paper?

(b) Explain a pictogram.

41. (a) Define an Ogive and how to obtain the value of median graphically.

(b) Illustrate graphically the distinction between a frequency polygon. A histogram and ogive curve. Comment on their uses.

(c) Define angular diagram. Discuss the usefulness of diagrammatic representation of facts.

(d) Explain and illustrate a Histogram.

42. Draw a rectangular diagram to represent the following information:

	Factory A	*Factory B*
Price per unit	Rs. 15.00	Rs. 12.00
Units produced	1,000 Nos.	1,200 Nos.
Raw material/unit	Rs. 5.00	Rs. 5.00
Other expenses/unit	Rs. 4.00	Rs. 3.00
Profit/unit	Rs. 6.00	Rs. 4.00

43. (a) Describe the utility of diagrammatic representation of statistical data.

(b) Following data relate to the expenditure of two families per month:

Items	*Family A*	*Family B*
Food	160	60
Rent	150	40
Clothing	100	30
Education	80	40
Lighting	30	10
Others	80	20

Represent the data by percentage bar diagram or pie chart.

44. Represent the following data relating to the expenditure of two families by means of a percentage bar diagram:

Expenditure items	*Family A Income Rs. 400*	*Family B Income Rs. 600*
Food	120	150
Clothing	80	100
Housing	60	100
Health & Education	40	80
Fuel & Lighting	40	40
Miscellaneous	40	60

45. Per capita incomes of 3 countries are given below:

Country	USA	UK	India
Per capita income (Rs.)	4900	3600	2500

Represent the data by means of circular diagram.

46. Prepare a histogram and a frequency polygon from the following data:

X :	0-10	10-20	20-30	30-40	40-60
f :	13	17	15	13	10

47. (a) Distinguish between statistics and parameter.

(b) Discuss the role of standard error in hypothesis testing.

(c) What do you mean by 'estimation'? Discuss the properties of an estimation. Differentiate between point and interval estimation.

(d) Distinguish between a null hypothesis and statistical hypothesis.

48.(a) (i) Explain the difference between statistics and parameter as used in sampling theory.

(ii) What do you understand by sampling distribution of a statistics and its standard error? Discuss the utility of standard error in statistics.

(b) (i) Explain the role of standard error in testing of a hypothesis.

(ii) "There is always a trade-off between Type I and Type II errors". Discuss.

(iii) Explain Type I and Type II errors in testing of a hypothesis. Also discuss the concept of level of significance.

(iv) Why should there be different formulae for testing the significance of difference in means when the samples are (i) small and (ii) large?

59. The following table gives details of monthly expenditures of 3 families A.B and C:

Items	*Average Monthly Expenditure (Rs.)*		
	A	*B*	*C*
Food	100	250	500
Clothing	20	40	60
Housing Rent	30	50	160
Fuel	20	50	100
Miscellaneous	30	110	180

Draw the percentage bar diagram.

50. Show the following data of expenditure of an average working class family by a suitable diagram:

Items of Expenditure	*% of Total Expenditure*
Food	65
Clothing	10
Housing	12
Fuel and Lighting	5
Miscellaneous	8

51. Given below is the pre-tax monthly income of residents of an industrial town:

Pre-tax income (Rs.)	*No. of Residents (in thousands)*
More than 7,000	2
More than 6,000	8
More than 5,000	10
More than 4,000	15
More than 3,000	35
More than 2,000	40
More than 1,000	55
More than 0	60

Draw a less than ogive cure and hence find out :

(i) the highest income of the lowest 50% of the residents: and

(ii) the minimum income earned by the top 5% of the residents.

52. Present the following data by a suitable diagram:

Items of Expenditure	*Family A*	*Family B*
Food	2000	2500
Clothing	1000	2000
House Rent	800	1000
Fuel and Lighting	400	500
Miscellaneous	800	2000
	5000	8000

53. (a) Construct a pie diagram' for the following frequency distribution:

Marks	:	10-19	20-29	30-39	40-49	50-59
No. of students	:	15	20	35	10	5

(b) The average highest and lowest prices of gold per 10 gm. in a country are given below. Show the data by using a range graph:

Years	*Minimum price (Rs.)*	*Maximum Price (Rs.)*
1990	1800	2000
1991	1850	2100
1992	1850	2050
1993	1900	2300
1994	1700	2100
1995	2000	2400
1996	2100	2500

54. Represent the following data by means of percentage subdivided bar diagram:

Cost per equipment	*1994 Rs.*	*1995 Rs.*	*1996 Rs.*
Raw materials	2160	2600	2700
Labour	540	700	810
Direct expenses	600	300	350
Factory expenses	360	200	300
Office expenses	180	200	270
Total	3840	4000	4490

55. (a) Discuss the methods of presentation of statistical data through graphs and diagrams.

(b) When do we use bar diagrams? How many different types of them are in use? Give some other forms of diagrammatic representations.

56. (a) Explain how tables, graphs and charts help in the effective presentation of data.

(b) What are the common methods for grouping and presentation of data?

(c) Explain the need and usefulness of diagrammatic representation of statistical data. What are the different types of diagram you know?

57. (a) Mention the advantages of graphic presentation of statistical data.

(b) What are the general rules for graphing the data?

58. As per Railway Budget how railways spend each rupee is given below. Draw a pie chart to represent the data. "Every one rupee the Railways spend, a hefty 3 paise is accounted for by staff wages and other allowances. 16 praise for depreciation reserve fund, 13 praise each for fuel and miscellaneous, 7 paise for dividend, 8 paise for stores and 23 paise for development works".

69. The following table gives the average monthly earnings of the mill workers in a certain city:

Monthly earnings (Rs.)	*No. of workers*	*Monthly earnings (Rs.)*	*No. of workers*
800-850	21	1200-1250	36
850-900	29	1250-1300	45
900-950	19	1300-1350	27
950-1000	39	1350-1400	48
1000-1050	43	1400-1550	21
1050-1100	94	1450-1500	12
1100-1150	73	1500-1550	9
1150-1200			

(i) Draw ogives by less than and more than methods for the data given below.

(ii) Find the number of workers whose wages lie between Rs. 1.180 and Rs. 1,480.

60. (a) What are the advantages of Graphic Presentation of Data of data?

(b) The merits of diagrammatic representation of data are classified under three main headings: attraction, effective impression and comparison. Explain and illustrate these points.

3

MEASURE OF CENTRAL VALUE

List of Formulae

Individual Series	*Desecrate Series*	*Continuous Series*
Arithmetic Mean		
Direct Method:	Direct Method	Direct Method
$\overline{X} = \frac{\Sigma X}{N}$	$\overline{X} = \frac{\Sigma fX}{N}$	$\overline{X} = \frac{\Sigma fm}{N}$
Short-cut Method	Short-cut Method	Short-cut Method
$\overline{X} = A + \frac{\Sigma X}{N}$	$\overline{X} = A + \frac{\Sigma fX}{N}$	$\overline{X} = A + \frac{\Sigma fm}{N}$
Step Deviation Method	Stem Deviation Method	
$\overline{X} = A + \frac{\Sigma X}{N} \times i$	$\overline{X} = A + \frac{\Sigma fm}{N} \times i$	
Median		
Size of $\frac{N+1}{2}$ th item	Size of $\frac{N+1}{2}$ th item	Size of $\frac{N+1}{2}$ th item
		$\text{Med.} = L + \frac{N/2 - \text{c.f.}}{f} \times i$
Mode		
Either by inspection or	$M_0 = L + \frac{\Delta_1}{\Delta_1 + \Delta_2} \times i$	
The value that occurs largest number of times.	grouping method determine that value around which most of the frefrequencies are concentrated.	$\Delta_1 = \lvert f_1 - f_2 \rvert$ $\Delta_1 = \lvert f'_1 - f'_2 \rvert$

Empirical Mode: Mode = 3 Median – 2 Mean

Geometric Mean		
$\text{G.M.} = \text{AL}\left(\frac{\Sigma \log X}{N}\right)$	$\text{G.M.} = \text{AL}\left(\frac{\Sigma f \log X}{N}\right)$	$\text{G.M.} = \text{AL}\left(\frac{\Sigma f \log m}{N}\right)$

Harmonic Mean

$$H.M. = AL\ \frac{N}{\Sigma(1/x)} \qquad H.M. = AL\ \frac{N}{\Sigma(f/x)} \qquad H.M. = \frac{N}{\Sigma(f/x)}$$

Weighted Arithmetic Mean	Weighted Geometric Mean	Weighted Harmonic Mean
$\overline{X}_w = \frac{\Sigma WX}{\Sigma W}$	$G.M._w = AL\left[\frac{\Sigma(\log XW)}{\Sigma W}\right]$	$H.M._w = \frac{\Sigma W}{\left(\frac{1}{a}\times W_1\right)+\left(\frac{1}{b}\times W_2\right)}$

Empirical Mode: Mode = 3 Median – 2 Mean

Combined Arithmetic Mean	Weighted Geometric Mean
$\overline{X}_{12} = \frac{N_1\overline{X}_1 + N_2\overline{X}_2}{N_1 + N_2}$	$G.M._{12} = AL \left(\frac{N_1 \log GM_1 + N_1 \log GM_2}{N_1 + N_2}\right)$

INTRODUCTION

One of the most important objectives of statistical analysis is to get one single value that describes the characteristic of the entire mass of unwieldy data. Such a value is called the central value or an 'average' or the expected value of the variable. The word average is very commonly used in day-to-day conversation. For example, we often talk of average boy in a class, average height or life of an Indian, average income, etc. When we say 'he is an average student' what it means is that he is neither very good nor very bad, just a mediocre type of student. However, in statistics the term average has a different meaning.

AVERAGE DEFINED

The word 'average' has been defined differently by various authors. Some important definitions are given below:

(a) "An average value is a single value within the range of the data that is used to represent all of the values in the series. Since an average is somewhere within the range of the data, it is also called a measure of central value." *–Croxton & Cowden*

(b) "The average is sometimes described as a number which is typical of the whole group." *–Leabo*

(c) "Average is an attempt to find one single figure to describe whole of figures." *–Clark*

It is clear from the above definitions that an average is a single value that represents a group of values. Such a value is of great significance because it depicts the characteristic of the whole group. Since an average represents the entire data. Its value lies somewhere in between the two extremes, *i.e.,* the largest and the smallest items. For this reason an average is frequently referred to as a measure of central tendency.

(d) "An average is a typical value in the sense that it is sometimes employed to represent all the individual values in a series or of a variable." *–Ya-Lun-Chou*

(e) "An average is a single value selected from a group of values to represent them in some way—a value which is supposed to stand for whole group, of which it is a part, as typical of all the values in the group." *–A.E. Waugh*

Objectives of Averaging

There are two main objectives of the study of averages:

(i) *To Facilitate Comparison Measure of central value,* by reducing the mass of data to one single figure, enable comparison to be made. Comparison can be made either at a point of time or over a period of time. For example, we can compare the percentage results of the students of different colleges in a certain examination, say. B. Com. for 1999, and thereby conclude which college is the best or we can compare the pass percentage of the some college for different time periods and thereby conclude as to whether the results are improving or deteriorating. Such comparisons are of immense help in framing suitable and timely policies. For example, if the pass percentage of students in College A in B. Com. was 80 in 1998 and 75 in 1999, the authorities have sufficient reason for investigating the possible cause of the deterioration in results.

However, while making comparison one should also take into consideration the multiplicity of forces that might be affecting the data. For example, if per capita income is rising in absolute terms from one period to another, it should not lead one to think that the standard of living is necessarily improving because the prices might be rising faster than the rise in per capita income and so in real terms people might be worse off. Moreover, the same measure should be used for making comparison between two or more groups. For example, we should not compare the mean wage of one factory with the median wage of another factory for drawing any inference about wage levels.

(ii) *To get single value that describes the characteristic of the entire group.* Measure of central value, by condensing the mass of data in one single value, enable us to get a bird's-eye view of the entire data. Thus one value can represent thousands, lakhs and even millions of values. For example, it is impossible to remember the individual incomes of millions of earning people of India and even if one could do it there is hardly any use. But if the average income is obtained by dividing the total national income by total population we get one single value that represents the entire population. Such a figure would throw light on the standard of living of an average Indian.

Requisites of a Good Average

Since an average is a single value representing a group of values, it is desired that such a value satisfies the following properties:

(i) *Based on all the Items :* The average should depend upon each and every item of the series so that if any of the items is dropped the average itself is altered. For example, the arithmetic mean of 10, 20, 30, 40, 50, is, $\frac{10 + 20 + 30 + 40 + 50}{5} = \frac{150}{5} = 30$. If we drop one item, say, 50, the arithmetic mean would be = 25.

(ii) *Rigidly Defined :* An average should be properly defined so that it has one and only one interpretation. It should preferably be defined by algebraic formula so that if different people compute the average from the same figures they all get the same answer (barring arithmetical mistakes). The average should not depend upon the personal prejudice and bias of the investigator, otherwise the results can be misleading.

(iii) *Not be unduly Affected by Extreme Observations :* Although each and every item should influence the value of the average, none of the items should influence it unduly. If one or two very small or very large items unduly affect the average, *i.e.*, either increase its value or reduce its value, the average cannot be really typical of the entire series. In other words, extremes may distort the average and reduce its usefulness.

(iv) *Simple to Compute :* An average should not only be easy to understand but also simple to compute so that it can be used widely. However, though ease of computation is desirable, it should not be sought at the expense of other advantages, *i.e.*, if in the interest of greater accuracy, use of a more difficult average is desirable, one should prefer that.

(v) *Easy to Understand :* Since statistical methods are designed to simplify complexity. It is desirable that an average be such that can be readily understood; otherwise, its use is bound to be very limited.

(vi) *Sampling Stability :* Last, but not the least, we should prefer to get a value which has what the statisticians call 'sampling stability'. This means that if we pick 10 different groups of college students, and compute the average of each group wc should expect to get approximately the same value. It does not mean, however, that there to get approximately the same value. It does not mean, however, that there can be no difference in the values of different samples. There may be some difference but those samples in which this difference (technically called sampling fluctuation) is less are considered better than those in which this difference is more.

(vii) *Capable of Further Algebraic Treatment :* We should prefer to have an average that could be used for further statistical computations so that its utility is enhanced. For example, if we are given the data about the average income and number of employees of two or more factories, we should be able to compute the combined average.

TYPES OF AVERAGES

The following are the important types of averages:

(a) Arithmetic mean :

(i) simple, and (ii) weighted.

(b) Geometric mean

(c) Harmonic mean

(d) Mode

(e) Median

Besides these, there are less important averages like moving average, progressive average, etc. These average have a very limited field of application and are, therefore, not so popular.

ARITHMETIC MEAN

The most popular and widely used measure of representing the entire data by one value is what most laymen call an 'average' and what the statisticians call the arithmetic mean. Its value is obtained by adding together all the items and by dividing this total by the number of items. Arithmetic mean may either be

(i) simple arithmetic mean, or (ii) weighted arithmetic mean.

Calculation of Simple Arithmetic Mean—Individual Observations

The process of computing mean in case of individual observations (*i.e.*, where frequencies are not given) is very simple. Add together the various values of the variable and divide the total by the number of items. Symbolically:

$$\overline{X}* = \frac{X_1 + X_2 + X_3 + \ldots + X_n}{N} \text{ or } \overline{X} = \frac{\Sigma X}{N}$$

Here $\overline{X}$ = Arithmetic Means, ΣX = Sum of all the values of the variable X, *i.e.*, X_1, X_2, X_3, ... X_n; N = Number of observations.

Steps. The formula involves two steps in calculating mean:

(i) Add together all the values of the variable X and obtain the total, *i.e.*, ΣX.

(ii) Divide this total by the number of the observations, *i.e.*, N.

Example 1:

The following table gives the monthly income of 10 employees in an office:

Income (Rs.)	*1,780*	*1,760*	*1,690*	*1,750*	*1,840*
	1,920	*1,100*	*1,810*	*1,050*	*1,950*

Calculate the arithmetic mean of incomes.

Let income be denoted by the symbol X.

Solution:

To calculate the arithmatic mean we have to construct the following table.

Calculation of Arithmetic Mean

Employee	***Monthly Income (Rs.)***	***Employee***	***Monthly Income (Rs.)***
1	1,780	6	1,920
2	1,760	7	1,100
3	1,690	8	1,810
4	1,750	9	1,050
5	1,840	10	1,950
		N = 10	ΣX = 16,650

Here, $\Sigma X = 16{,}650$, $N = 10$

$$\overline{X} = \frac{16{,}650}{10} = 1{,}665.$$

Hence the average income is Rs. 1,665.

Short-cut Method : We can calculate the arithmatic mean by short-cut method. When derivations are taken from an arbitrary origin, the formula for calculating arithmetic mean is

$$\overline{X} = A + \frac{\Sigma d}{N}$$

where A is the assumed mean and d is the deviation of items from assumed mean, *i.e.,* d = (X – A).

Steps.

(1) Take an assumed mean.

(2) Take the deviations of items from the assumed mean and denote these deviations by d.

(3) Obtain the sum of these deviations, *i.e.,* Σd.

(4) Apply the formula : $\overline{X} = A + \frac{\Sigma d}{N}$

Calculate arithmetic mean by taking 1,800 as the assumed mean.

Solution:

Calculation of Arithmetic Mean

Employee	*Income*	*(X – 1800)*
1	1,780	–20
2	1,760	–40
3	1,690	–110
4	1,750	–50
5	1,840	+40
6	1,920	+120
7	1,100	–700
8	1,810	+10
9	1,050	–750
10	1,950	+150
N = 60		Σd = –1350

$$\overline{X} = A + \frac{\Sigma d}{N}$$

A = 1,800, Σd = – 1,350, N = 10

$$\overline{X} = 1{,}800 - \frac{1{,}350}{10} = 1{,}800 - 135 - 1{,}665.$$

Example 2:

From the following data of the marks obtained by 60 students of a class, calculate the arithmetic mean:

Marks	*No. of Students*	*Marks*	*No. of Students*
20	*8*	*50*	*10*
30	*12*	*60*	*6*
40	*20*	*70*	*4*

Let the marks be denoted by X and the number of students by f.

Solution:

To calculate the arithmatic mean we have to construct the following table.

Calculation of Arithmetic Mean

Marks X	*No. of students* f	*fX*
20	8	160
30	12	360
40	20	800
50	10	500
60	6	360
70	4	280
	N = 60	ΣfX = 2,460

$$\overline{X} = \frac{\Sigma fX}{N} = \frac{2,460}{60} = 41$$

Hence the average marks = 41.

Calculation of Arithmetic Mean—Discrete Series

In discrete series arithmetic mean may be computed by applying

(i) Direct method, or

(ii) Short-cut method.

Direct Method

The formula for computing mean is

$$\overline{X} = \frac{\sum fX}{N}$$

where, f = Frequency; X = The variable in question;

N = Total number of observation, *i.e.*, $\sum f$.

Steps:

(i) Multiply the frequency of each row with the variable and obtain the total $\sum fX$.

(ii) Divide the total obtained by step (i) by the number of observations *i.e.*, total frequency.

Example 3:

From the following data compute arithmetic mean by direct method.

Marks	*0–10*	*10–20*	*20–30*	*30–40*	*40–50*	*50–60*
No. of students	*5*	*10*	*25*	*30*	*20*	*10*

Solution:

Calculation of Arithmetic Mean by Direct Method

Marks	***Mid-point*** ***m***	***No. of Students*** ***f***	***fm***
0–10	5	5	25
10–20	15	10	150
20–30	25	25	625
30–40	35	30	1,050
40–50	45	20	900
50–60	55	10	550
		N = 100	$\sum fm$ = 3,300

$$\overline{X} = \frac{\sum fm}{N} = \frac{3{,}300}{100} = 33.$$

Short-cut Method : When short-cut method is used, arithmetic mean is computed by applying the following formula;

$$\overline{X} = A + \frac{\sum fd}{N}$$

when A = assumed mean;

d = deviations of mid-points from assumed mean, *i.e.*, (m – A);

N = total number of observations.

Steps:

(i) Take an assumed mean.

(ii) From the mid-point of each class deduct the assumed mean.

(iii) Multiply the respective frequencies of each class by these deviations and obtain the total Σfd.

(iv) Apply the formula: $\overline{X} = A + \frac{\Sigma fd}{N}$

Calculate arithmetic mean by the short-cut method.

Solution:

Calculation of Arithmetic mean

Mark	*Mid-point* *m*	*No. of Students* *f*	*(m – 35)* *d*	*fd*
0–10	5	5	–30	–150
10–20	15	10	–20	–200
20–30	25	25	–10	–250
30–40	35	30	0	0
40–50	45	20	+10	+200
50–60	55	10	+20	+200
		N = 100		Σfd = –200

$$\overline{X} = A + \frac{\Sigma fd}{N} = 35 - \frac{200}{100} = 35 - 2 = 33$$

In order to simplify the calculations, we can divide the deviations by class intervals, *i.e.*, calculate (m – A)/i and then multiply by in the formula for getting mean. The formula becomes;

$$\overline{X} = A + \frac{\Sigma fd}{N} \times i$$

It may be pointed out that when class intervals are unequal we can simplify calculations by taking a common factor. In such a case we should use (m – A)/C instead (m – A)/i while making calculations.

Compute arithmetic mean by step deviation method.

Solution:

Calculation of Arithmetic Mean

Marks	*Mid-point* m	*No. of Students* f	*(m – 35)* d	*(m – 35)/10*	*fd*
0–10	5	5	–30	–3	–15
10–20	15	10	–20	–2	–20
20–30	25	25	–10	–1	–25
30–40	35	30	0	0	0
40–50	45	20	+10	+1	+20
50–60	55	10	+20	+2	+20
		N = 100			Σfd = –20

$$\overline{X} = A + \frac{\Sigma fd}{N} \times i = 35 - \frac{20}{100} \times 10 = 35 - 2 = 33.$$

It is clear from above that all the three methods of finding arithmetic mean in continuous series give us the same answer. The direct method, though the simplest, involves more calculations when mid-points and frequencies are very large in magnitude. For example, observe the following data:

Income in Rs.	*No. of Persons*
400–500	368
500–600	472
600–700	969
700–800	567
800–900	304

In this case step deviation method would be far simpler. In fact, step deviation method should be adopted wherever possible because it minimises the calculations.

While computing mean in continuous series the mid-points of the various classes are taken as representative of that particular class. The reason is that when the data are grouped, the exact frequency with which each value of the variable occurs in the distribution is unknown. We only know the limits within which a certain number of frequencies occur. For example, when we say that the number of persons within the income group 400–500 is 50 we cannot say as to how many persons out of 50 are getting 401, 402, 403, etc. We, therefore, make an assumption while calculating arithmetic mean that the frequencies within each class are spread evenly over the range of the class interval, *i.e.,* there will be as many items below the mid-point as above it. Unless such an assumption is made the value of mean cannot be computed.

This assumption is likely to led to some error. As a result thereof the mean of a number of observations calculated from a frequency distribution will generally be only an approximation to the mean calculated from the original data. However, the possibility of compensating errors must be considered. Some of the mid-points err by being too low and others err by being too high. In general, then mid-points of the classes below the class containing the arithmetic mean tend to be too low and the mid-points of the classes above the class containing the arithmetic mean tend to be too high. It is quite possible, therefore, that when the errors are assumed, those which are too low will offset, in part at least, those which are too high, so that the arithmetic mean for the entire distribution will be approximetely of the same value as is obtained from a list of values.

Example 4:

Calculate arithmetic mean by the short-cut method using frequency distribution.

Solution:

Calculation of Arithmetic Mean

Marks *X*	*No. of Students* *f*	*(X – 40)* *d*	*fd*
20	8	–20	–160
30	12	–10	–120
40	20	0	0
50	10	+10	+100
60	6	+20	+120
70	4	+30	+120
	N = 60		Σfd = 80

$$\overline{X} = \frac{\Sigma fX}{N} = 40 + \frac{60}{60} = 40 + 1 = 41$$

Calculation of Arithmetic Mean—Continuous Series

In continuous series, arithmetic mean may be computed by applying any of the following methods:

(i) Direct method

(ii) Short-cut method.

Direct Method : When direct method is used

$$\overline{X} = \frac{\Sigma fX}{N}$$

where m = mid-point of various classes; f = the frequency of each class;

N = the total frequency.

Steps:

(i) Obtain the mid-point of each class and denote it by m.

(ii) Multiply these mid-points by the respective frequency of each class and obtain the total Σfm.

(iii) Divide the total obtained in step (i) by the sum of the frequency, *i.e.,* N.

Example 5:

Calculate arithmetic mean from the following data:

Marks	*0–10*	*10–30*	*30–60*	*60–100*
Nos. of	*5*	*12*	*25*	*8*

Solution:

The class intervals are unequal but still to simplify calculations we can take 5 as the common factor.

Calculation of Mean

Marks	*Mid-point* *m*	*f*	*(m – 45)/5* *d*	*fd*
0–10	5	5	–8	–40
10–30	20	12	–5	–60
30–60	45	25	0	0
60–100	80	8	+7	+56
			N = 50	Σfd = –44

$$\overline{X} = A + \frac{\Sigma fd}{N} \times C$$

A = 45, Σfd = –44, N = 50, C = 5

$$\overline{X} = 45 - \frac{44}{50} \times 5 = 45 - 4.4 = 40.6$$

Correcting Incorrect Values. It sometimes happens that due to an oversight or mistake in copying, certain wrong items are taken while calculating mean. The problem is how to find out the correct mean. The process is very simple. From incorrect ΣX deduct wrong items and add correct items and then divide the correct ΣX by the number of observations. The result, so obtained, will give the value of correct mean.

Example 6:

Mean of 100 observations is found to be 40. If at the time of computation two items are wrongly taken as 30 and 27 instead of 3 and 72. Find correct Mean.

Solution:

$$\overline{X} = \frac{\Sigma X}{N} \quad \text{or} \quad \Sigma X = N\overline{X}$$

Here $\overline{X} = 40$, $N = 100$

$\therefore \Sigma X = 100 \times 40 \quad = \quad 4000$

Less incorrect items $= 57$

$= 3943$

Add correct items $= 75$

Correct total $= 4018$

Correct mean $= \frac{4018}{100} = 40.18.$

Calculation of Arithmetic Mean in Case of Open-end Classes

Open-end classes are those in which lower limit of the first class and the upper limit of the last class are known in such a case we cannot find out the arithmetic mean unless we make an assumption about the unknown limits. The assumption would naturally depend upon the class interval following the first class and preceding the last class. For example, observe the following data:

Marks	*No. of students*	*Marks*	*No. of students*
Below 10	4	30–40	15
10–20	6	40–50	8
20–30	10	Above 50	7

In the above case since the class interval is uniform, the appropriated assumption would be that the lower limit of the first class is zero and the upper limit of the last class is 60. The first class thus would be 0–10 and the last class 50–60. Observe another case:

Marks	*No. of students*	*Marks*	*No. of students*
Below 10	4	60–100	7
10–30	6	Above 100	3
30–60	10		

In the above case since the class interval is 20 in the second class, 20 in the third class, 40 in the fourth class, *i.e.,* it is increasing by 10. The appropriate assumption would be that the lower limit of the first class is zero and the upper limit of the last class 150. In other words, first class is 0–10 and the lase one 100–150.

If the class intervals are of varying width, an effort should not be made to determine the lower limit of the lowest class and upper limit of the highest class. The use of median or mode would be better in such a case. Because of the difficulty of ascertaining lower limit and upper limit in open-end distributions it is suggested that in such distributions arithmetic mean should not be used.

Mathematical Properties of Arithmetic Mean

The following are a few important mathematical properties of the arithmetic mean:

1. The sum of the deviations of the items from the arithmetic mean (taking signs into account) is always zero, *i.e.,* $\Sigma(X - \overline{X}) = 0$. This would be clear from the following example:

X	$(X - \overline{X})$
10	– 20
20	– 10
30	0
40	+ 10
50	+ 20
$\Sigma X = 150$	$\Sigma(X - \overline{X}) = 0$

Here $\overline{X} = \frac{\Sigma X}{N} = \frac{150}{5} = 30$. When the sum of the deviations from the actual mean, *i.e.,* 30, is taken it comes out to be zero. It is because of this property that the mean is characterised as point of balance, *i.e.,* the sum of the positive deviations from it is equal to the sum of the negative deviations from it.

2. The sum of the squared deviations of the items from arithmetic mean is minimum, that is, less than the sum of the squared deviations of the items from any other value. The following example would verify the point:

X	$(X - \overline{X})$	$(X - 4)^2$
2	–2	4
3	–1	1
4	0	0
5	+1	1
6	+2	4
$\Sigma X = 20$	$\Sigma(X - \overline{X}) = 0$	$\Sigma(X - \overline{X})^2 = 10$

The sum of the squared deviations is equal to 10 in the above case. If the deviations are taken from any other value the sum of the squared deviations would be greater than 10. For example, let us calculate the squares of the deviations of item from a value less than the arithmetic mean, say 3.

X	$(X - 3)$	$(X - 3)^2$
2	– 1	1
3	0	0
4	+ 1	1
5	+ 2	4
6	+ 3,	9
		$\Sigma(X - 3)^2 = 15$

It is clear that $(\Sigma X - \overline{X})^2$ is greater. This property that the sum of the squares of items is least from the means is of immense use in regression analysis which shall be discussed later.

3. Since $\overline{X} = \dfrac{\Sigma X}{N}, \; N\overline{X} = \Sigma X$

In other words, if we replace each item in the series by the mean, then the sum of these substitutions will be equal to the sum of the individual items. For example, in the discussion of first property $\Sigma X = 150$ and the arithmetic mean 30. If for each item we substitute 30, we get the same total, *i.e.*, $30 + 30 + 30 + 30 + 30 = 150$.

This property is of great practical value. For example, if we know the average wage in a factory, say, Rs. 1,060 and the number of workers employed, say, 200, we can compute total wages bill from the relation $\frac{N\overline{X} = \Sigma X}{X.0}$. The total wage bill in this case would be 200 × 1,060, *i.e.*, Rs. 2,12,000 which is equal to ΣX.

4. If we have the arithmetic mean and number of items of two or more than two related groups, we can compute combined average of these groups by applying the following formula:

$$\overline{X}_{12} = \frac{N_1 \overline{X}_1 = \overline{X}_2}{N_1 + N_2}$$

$\overline{X}_{12}$ = combined mean of the two groups

$\overline{X}_1$ = arithmetic mean of first group

$\overline{X}_2$ = arithmetic mean of second group

N_1 = number of items in the first group

N_2 = number of items in the second group

Example 7:

The mean marks of 190 students were found to be 40. Later on it was discovered that a score of 53 was misread as 83. Find the correct mean corresponding to the correct score.

Solution:

We are given $N = 100, \quad \overline{X} = 40$

Since $\overline{X} = \frac{\Sigma X}{N}$

$$\Sigma X = N\overline{X} = 100 \times 40 = 4000$$

But this is not correct ΣX

Correct ΣX = Incorrect ΣX – wrong item + correct item

$= 4000 - 83 + 53 = 3970$

$\therefore$ Correct $\overline{X} = \frac{\text{correct } \Sigma X}{N} = \frac{3970}{100} = 39.7$

Hence the correct average = 39.7

Example 8:

The mean height of 25 male workers in a factory is 61 cm. and the mean height of 35 female workers in the same factory is 58 cm. Find the combined mean height of 60 workers in the factory.

Solution:

$$\overline{X}_{12} = \frac{N_1 \overline{X}_1 = \overline{X}_2}{N_1 + N_2}$$

$$N_1 = 25,\ \overline{X}_1 = 61,\ N_2 = 35,\ \overline{X}_2 = 58$$

$$\overline{X}_{12} = \frac{(25 \times 61) + (35 \times 58)}{25 + 35} = \frac{1525 + 2030}{60} = \frac{3555}{60} = 59.25$$

Thus the combined mean height of 60 workers is 59.25 cm.

If we have to find out the combined mean of three sub-groups the above formula can be extended as follows:

$$\overline{X}_{123} = \frac{N_1 \overline{X}_1 + N_2 \overline{X}_2 + N_3 \overline{X}_3}{N_1 + N_2 + N_3}.$$

MERITS AND LIMITATIONS OF ARITHMETIC MEAN

Merits

Arithmetic mean is most widely used in practice because of the following reasons:

(a) Being determined by a rigid formula, if lends itself to subsequent algebraic treatment better than the median or mode.

(b) It is defined by a rigid mathematical formula with the result that everyone who computes the average gets the same answer.

(c) It is affected by the value of every item in the series.

(d) It is the simplest average to understand and easiest to compute. Neither the arraying of data as required for calculating median for grouping of data as required for calculating mode is needed while calculating mean.

(e) The mean is typical in the sense that it is the centre of gravity, balancing the values on either side of it.

(f) It is a calculated value, and not based on position in the series.

(g) It is relatively reliable in the sense that it does not vary too much when repeated samples are taken from one and the same population, at least not as much as some other kind of statistical descriptions.

Limitations

Since the value of mean depends upon each and every item of the series, extreme items, *i.e.*, vary small and very large items, unduly affect the valuc of the average. For example, if in a tutorial group there are 4 students and their marks in a test are 60, 70, 10 and 80 the average marks would be $\frac{60 + 70 + 10 + 80}{4} = \frac{220}{4} = 55$. One single item, *i.e.*, 10, has reduces the average marks considerably. The smaller the number of observations, the greater is likely to be the impact of extreme value.

(a) The arithmetic mean is not always a good measure of central tendency. The mean provides a "characteristic" value, in the sense of indicating where most of the values lie, only when the distribution of the variable is reasonably normal (bell-shaped). In case of a U-shaped distribution the mean is not likely to serve a useful purpose.

(b) In a distribution with open-end classes the value of mean cannot be computed without making assumptions regarding the size of the class interval of the open-end classes. If such classes contain a large proportion of the values, then mean may be subject to substantial error. However, the values of the median and mode cane computed where there open-end classes without making any assumptions about size of class interval.

Weighted Arithmetic Mean

One of the limitations of the arithmetic mean discussed above is that it gives equal importance to all the items. But there re cases where the relative importance of the different items is not the same. When this is so, we compute weighted arithmetic mean. The term 'weight' stands for the relative importance of the different items. The formula for computing weighted arithmetic mean is:

$$\overline{X}_{\omega} = \frac{\Sigma WX}{\Sigma W}$$

where $\overline{X}_{\omega}$ represents the weighted arithmetic mean; X represents the variable values, *i.e.*, $X_1, X_2, ..., X_n$.

We represents the weights attached to variable values, *i.e.*, $\omega_1, \omega_2, ... \omega_n$, respectively.

(i) Multiply the weights by the variable X and obtain the total ΣWX.

(ii) Divide this total by the sum of the weights, *i.e.*, ΣW.

In case of frequency distribution, if $f_1, f_2, \ldots f_n$ are the frequencies of the variable values $X_1, X_2, \ldots, X_n$ respectively then the weighted arithmetic mean is given by:

$$\overline{X}_\omega = \frac{\Sigma W(fX)}{\Sigma W}$$

From the expanded form

$$\overline{X}_\omega = \frac{W_1 (f_1 X_1) + W_2 (f_2 X_2) + \ldots + W_n (f_n X_n)}{W_1 + W_2 + \ldots + W_n}.$$

An important problem that arises while using weighted mean is regarding selection of weights. Weights may be either actual or arbitrary, *i.e.*, estimated. Needless to say, if actual weights are available, nothing like this. However, in the absence of actual weights, arbitrary or imaginary weights may be used. The use of arbitrary weights may lead to some error, but it is better than no weights at all. In practice. It is found that if weights are logically assigned keeping the phenomena in view, the error involved will be so small that it can be easily over looked.

It should be noted that:

(i) Simple arithmetic mean shall be equal to the weighted arithmetic mean if the weights are equal. Symbolically,

$$\overline{X} = \overline{X}_\omega \text{ if } W_1 = W_2.$$

(ii) Simple arithmetic mean shall be less than the weighted arithmetic mean if and only if greater weights are assigned to greater values and smaller weights are assigned to smaller values. Symbolically,

$$\overline{X} < \overline{X}_\omega \text{ if } (\omega_2 - \omega_1)(X_1 - X_2) < 0.$$

(iii) Simple arithmetic mean is greater than the weighted arithmetic mean if and only if smaller weights is attached to the higher values and greater weight is attached to the smaller values. Symbolically.

$$\overline{X} > \overline{X}_\omega \text{ if } (w_2 - w_1)(X_1 - X_2) < 0.$$

It may be noted that weighted arithmetic mean is specially useful in problems relating to:

(i) Construction of index numbers, and

(ii) Standardized birth and death rates.

Example 9:

Comment on the performance of the students of the three universities given below using simple and weighted averages:

University Course study	*Pass %*	*Bombay No. of Students (in hundreds)*	*Pass %*	*Calcutta No. of Students (in hundreds)*	*Pass %*	*Madras No. of Students (in hundreds)*
M.A.	*71*	*3*	*82*	*2*	*81*	*2*
M.Com.	*83*	*4*	*76*	*3*	*76*	*3.5*
B.A.	*73*	*5*	*73*	*6*	*74*	*4.5*
B.Com.	*74*	*2*	*76*	*7*	*58*	*2*
B.Sc.	*65*	*3*	*65*	*3*	*70*	*7*
M.Sc.	*66*	*3*	*60*	*7*	*73*	*2*

Solution:

University of Study	*Pass %*	*Bombay No. of Students (in hundreds)*		*Pass %*	*Calcutta No. of Students (in hundreds)*		*Pass %*	*Madras No. of Students (in hundreds)*	
	X	W	WX	X	W	WX	X	W	WX
M.A.	71	3	213	82	2	164	81	2.0	162
M.Com.	83	4	332	76	3	228	76	3.5	266
B.A.	73	5	365	73	6	438	74	4.5	333
B.Com.	74	2	148	76	7	532	58	2.0	116
B.Sc.	65	3	195	65	3	195	70	7.0	490
M.Sc.	66	3	198	60	7	420	73	2.0	146
	ΣX = 432	ΣW = 20	ΣWX = 1,451	ΣX = 432	ΣW = 28	ΣWX = 1,977	ΣX = 432	ΣW = 21	ΣWX = 1,51

Simple and Weighted Arithmetic Mean

Bombay $\overline{X} = \frac{\Sigma X}{N} = \frac{432}{6} = 72;\qquad \overline{X}_{\omega} = \frac{\Sigma WX}{\Sigma W} = \frac{1{,}451}{20} = 72.55$

Calcutta $\overline{X} = \frac{\Sigma X}{N} = \frac{432}{6} = 72;\qquad \overline{X}_{\omega} = \frac{\Sigma WX}{\Sigma W} = \frac{1{,}977}{28} = 70.61$

Madras $\overline{X} = \frac{\Sigma X}{N} = \frac{432}{6} = 72;\qquad \overline{X}_{\omega} = \frac{\Sigma WX}{\Sigma W} = \frac{1{,}513}{21} = 72.05$

The arithmetic mean is the same for all the three universities, *i.e.,* 72 and hence, it may be concluded that the performance of students is alike.

But this will be a wrong conclusion because what we should compare here is the weighted arithmetic mean. On comparing the weighted arithmetic means we find that for Bombay the meàn value is the highest and hence we can say that in Bombay University the performance of students is best.

Median

As distinct from the arithmetic mean which is calculated from the value of every item in the series, the median is what is called a positional average. The term 'position' refers to the place of a value in a series. The place of the median in a series is such that an equal number of items lie on either side of it.

For example, if the income of five employees is Rs. 900, 950, 1020, 1200 and 1280 the median would be 1020.

900

950

1020 value at middle position of the array

1200 } ← there are two middle positin values
1280

For the above example the calculation of median was simple because of odd number of observations. When an even number of observations are listed, there is no single middle position value and the median is taken to be the arithmetic mean of two middle most items. For example, if in the above case we are given the income of six employees as 900, 950, 1020, 1200, 1280, 1300, the median income would be;

900

950

1020

1200

1280

1300

$$\text{Median} = \frac{1020 + 1200}{2} = \frac{2220}{2} = 1110.$$

Hence, in case of even number of observations median may be found by averaging two middle position values.

Thus, when N is odd, the median is an actual value, with the remainder of the series in two equal parts on either side of it. If N is even, the median is a derived figure, *i.e.*, half the sum of the middle values.

Calculation of Median—Individual Observations

(i) Arrange the data in ascending or descending order of magnitude. (Both arrangements would give the same answer.)

(ii) In a group composed of an odd number of values such as 7, add 1 to the total number of values and divide by 2. Thus, 7 + 1 would be 8 which divided by 2 gives 4—the number of the values starting at either and of the numerically arranged groups will be the median value. In a large group the same method may be followed. In a group of 199 items the middle value would be 100th value. This would be determined by $\frac{199+1}{2}$. In the form of formula:

$$\text{Med.} = \text{Size of } \frac{N+1}{2}\text{th item.}$$

Example 10:

A train runs 25 miles at a speed of 30 m.p.h., another 50 miles at a speed of 40 m.p.h., then due to repairs of the track travels for 6 minutes at a speed of 10 m.p.h. and finally covers the remaining distance of 24 miles at a speed of 24 m.p.h. What is the average speed in miles per hour?

Solution:

Time taken in covering 25 miles at a speed of 30 m.p.h. = 50 minutes. Time taken in covering 50 miles at a speed of 40 m.p.h. = 75 minutes. Distance covered in 6 minutes at a speed of 10 m.p.h. = 1 mile. Time taken in covering 24 miles at a speed of 24 m.p.h. = 60 minutes.

Therefore, taking the time taken as weights we have the weighted mean as

Speed in m.p.h. *XW*	*Time taken* *W*	*WX*
30	50	1,500
40	75	3,000
10	6	60
24	60	1,440
		ΣWX = 6,000

$$\therefore \text{Average speed} = \frac{6{,}000}{191} = 31.41 \text{ m.p.h.}$$

Example 11

A contractor employs three types of workers–male, female and children. To a male he pays Rs. 40 per day, to a female worker Rs. 32 per day and to a child worker Rs. 15 per day. What is the average wage per day paid by the contractor?

Solution:

The average wage is not the simple arithmetic mean, i.e, $\frac{40 + 32 + 15}{3}$ = Rs. 29 per day. It we assume that the number of male, female and child workers is the same, this answer would be correct. For example, if we take 10 workers in each case then the mean wage would be

$$\frac{(10 \times 40) + (10 \times 32) + (10 \times 15)}{10 + 10 + 10} = \frac{400 + 320 + 150}{30} = \text{Rs. } 29$$

However, the number of male, female and child workers employed is generally different, it we know how many workers of each type are employed by the contractor in question, nothing like this. However, in the absence of this we take assumed weights. Let as assume that the number of male, female and child workers employed is 20, 15 and 5 respectively. The average wage would be the weighted mean calculated as follows:

Wages per day (Rs.) ***XW***	***No. of workers*** ***W***	***WX***
40	20	800
32	15	480
15	5	75
	SW = 40	ΣWX = 1,355

$$\overline{X}_{\omega} = \frac{\Sigma WX}{\Sigma W} = \frac{1{,}355}{40} = 33.875 \text{ or } 33.88$$

Example 12:

Calculate the median for the following frequency distribution:

Marks	***No. of Students***	***Marks***	***No. of Students***
45–50	*10*	*20–25*	*31*
40–45	*15*	*15–20*	*24*
35–40	*26*	*10–15*	*15*
30–35	*30*	*5–10*	*17*
25–30	*42*		

Solution:

First arrange the data in ascending order and then find out median.

Calculation of Median

Marks	*f*	*c.f.*	*Marks*	*f*	*c.f.*
5–10	7	7	30–35	30	149
10–15	15	22	35–40	26	175
15–20	24	46	40–45	15	190
20–25	31	77	45–50	10	200
25–30	42	119			

$$\text{Med.} = \text{size of } \frac{N}{2} \text{ item} = \frac{200}{2} = 100\text{th item}$$

Median lies in the class 25–30

$$\text{Med.} = L + \frac{N/2 - c.f.}{f} \times i$$

$$L = 25,\ N/2 = 100,$$

$$c.f. = 77,\ f = 42,\ i = 5$$

$$\text{Med. } 25 + \frac{100 - 77}{42} \times 5 = 25 + 2.74 = 27.74.$$

Example 13:

Obtain the value of median from the following data:

391 384 591 407 672

522 777 753 2,488 1,490

Solution:

Calculation of Median

Sl. No.	*Data arranged in ascending order (X)*	*Sl. No.*	*Data arranged in ascending order (X)*
1	384	6	672
2	391	7	753
3	405	8	777
4	522	9	1,490
5	591	10	2,488

$$\text{Median} = \text{size of } \frac{N+1}{2}\text{th item} = \frac{11}{2} = 5.5\text{th item.}$$

$$\text{Size of 5.5th item} = \frac{\text{5th item} + \text{6th item}}{2} = \frac{591 + 672}{2} = \frac{1{,}263}{2} = 631.5$$

Computation of Median—Discrete Series

Steps:

(i) Arrange the data in ascending or descending order of magnitude.

(ii) Find out the cumulative frequencies.

(iii) Apply the formula: Median = size of $\frac{N+1}{2}$.

(iv) Now look at the cumulative frequency column and find that total which is either equal to $\frac{N+1}{2}$ or next higher to that and determine the value of the variable corresponding to it. That gives the value of median.

Example 14:

From the following data of the wages of 7 workers compute the median wage:

Wages (in Rs.) *1100* *1150* *1080* *1120* *1200* *1160* *1400*

Solution:

Calculation of Median

Sl. No.	*Wages arranged in ascending order*	*Sl. No.*	*Wages arranged in ascending order*
1	1080	5	1160
2	1100	6	1200
3	1120	7	1400
4	1150		

$$\text{Median} = \text{size of } \frac{N+1}{2}\text{th item} = \frac{7+1}{2} = 4\text{th item.}$$

Size of 4th item = 1150. Hence the median wage = Rs. 1150.

We thus find that median is the middlemost item : 3 persons get a wage less than Rs. 1150 and equal number, *i.e.*, 3, get more than Rs. 1150.

The procedure for determining the median of an even-numbered group of items is not as obvious as above. If there were, for instance, different values in a group, the median is really not determinable since both the 5th and 6th values are in the centre. In practice, the median value for group composed of an even number of items is estimated by finding the arithmetic mean of the two middle values—that is, adding the two values in the middle and dividing by two. Expressed in the form of formula, it amounts to:

$$\text{Median} = \text{Size of } \frac{N+1}{2}\text{th item}$$

Thus, we find that it is both when N is odd as well as even that 1 (one) has to be added to determine median value.

Example 15:

From the following data find the value of median:

Income (Rs.)	*1000*	*1500*	*800*	*2000*	*2500*	*1800*
No. of persons	*24*	*26*	*16*	*20*	*6*	*30.*

Solution:

Calculation of Median

Income arranged in ascending order	***No. of persons f***	***c.f.***	***Income arranged in ascending order***	***No. of persons f***	***c.f.***
800	16	16	1800	30	96
1000	24	40	2000	20	116
1500	26	66	2500	6	122

$$\text{Median} = \text{Size of } \frac{N+1}{2}\text{th item} = \frac{122+1}{2} = 61.5\text{th item.}$$

Size of 61.5th item = 1500.

Calculation of Median—Continuous Series

Steps. Determine the particular class in which the value of median lies. Use N/2 as the rank of the median and not (N + 1)/2. Some writers have suggested that while calculating median in continuous series 1 should be added to total frequency if it is odd (say 99) and should not be added if it is even figure (say, 100). However, 1 is to be added in case of individual and discrete series because specific items and individual values are involved.

In a continuous frequency distribution all the frequencies lose their individuality. The effort now is not to find the value of one specific item but to find a particular point on a curve—that one value which will have 50 percent of frequencies on one side of it and 50 percent of the frequencies on the other. It will be wrong to use the above rule. Hence it is N/2 which will divide the area of curve into two equal parts and as such we should use N/2 instead of (N + 1)/2, in continuous series. After ascertaining the class in which median lies, the following formula is used for determining the exact value of median.

$$\text{Median} = L + \frac{N/2 - c.f.}{f} \times i$$

L = Lower limit of the median class, *i.e.*, the class in which the middle item of the distribution lies.

c.f. = Cumulative frequency of the class preceding the median class or sum of the frequencies of all classes lower than the median class.

f = Simple frequency of the median class.

i = The class interval of the median class.

It should be remembered that while interpolating the median value in a frequency distribution it is assumed that the variable is continuous and that there is an orderly and even distribution of items within each class.

Example 16:

Calculate the median from the following data:

Weight (in gms.)	*No. of Apples*	*Weight (in gms.)*	*No. of Apples*
410–419	*14*	*450–459*	*45*
420–429	*20*	*460–469*	*18*
430–439	*42*	*470–479*	*7*
440–449	*54*		

Solution:

Since we are given inclusive class intervals, we should convert in to the exclusive the by deducting 0.5 from the lower limits and adding 0.5 to the upper limits.

Weight	*f*	*c.f.*
409.5–419.5	14	14
419.5–429.5	20	34
429.5–439.5	42	76
439.5–449.5	54	130
449.5–459.5	45	175
459.5–469.5	18	193
469.5–479.5	7	200
	N = 200	

$$\text{Med.} = \text{Size of } \frac{N}{2}\text{th item} = \frac{200}{2} = 100\text{th item}$$

Median lies in the class 439.5 – 449.5

$$\text{Med.} = L + \frac{N/2 - c.f.}{f} \times i$$

$L = 439.5$, $N/2 = 100$,

$c.f. = 76$, $f = 54$, $i = 10$

$$\text{Med.} = 439.5 + \frac{100 - 76}{54} \times 10$$

$$= 439.5 + 4.44 = 443.94.$$

Example 17:

Compute median from the following data:

Mid-value	*Frequency*	*Mid-value*	*Frequency*
115	*6*	*165*	*60*
125	*25*	*175*	*38*
135	*48*	*185*	*22*
145	*72*	*195*	*3*
155	*116*		

Solution:

Since we are given the mid-values, we should find out the upper and lower limits the various classes.

Calculation of Median

Class group	*f*	*c.f.*	*Class group*	*f*	*c.f.*
100–120	6	6	160–170	60	327
120–130	25	31	170–180	38	365
130–140	48	79	180–190	22	387
140–150	72	151	190–200	3	390
150–160	116	267			

$$\text{Med.} = \text{Size of } \frac{N}{2}\text{th item} = \text{Size of } \frac{390}{2} = 195\text{th item}$$

Median lies in the class 150–160

$$\text{Median} = L + \frac{N/2 - c.f.}{f} \times i$$

$$L = 150,\ N/2 = 195,$$

$$c.f. = 151,\ f = 116,\ i = 10$$

$$\text{Median} = 150 + \frac{195 - 151}{116} \times 10 = 150 + 3.79 = 153.79.$$

Example 18:

An incomplete distribution is given below:

Variable:	*0–10*	*10–20*	*20–30*	*30–40*	*40–50*	*50–60*	*60–70*
Frequency:	*10*	*20*	*?*	*40*	*?*	*25*	*15*

(i) You are given that the median value is 35. Find out missing frequency (given the total frequency = 170)

(ii) Calculate the arithmetic mean of the completed table.

Solution:

Let the missing frequency of the class 20–30 be denoted by f_1 and that of 40–50 by f.

The total frequency = 170

The frequencies of the classes other than the missing ones are (10 + 20 + 40 + 25 + 15) = 110.

$$110 + f_1 + f_2 = 170$$

Hence $f_1 + f_2 = (170 - 110) = 60$

$$\text{Med.} = L + \frac{N/2 - c.f.}{f} \times i$$

$$\text{Med.} = \text{Size of } \frac{N}{2} \text{th item} = \frac{170}{2} = 85\text{th item}$$

We are given median = 35

Hence it must lie in the class 30–40.

Thus the various values known to us are

Med. = 35, L = 30, N/2 = 85, c.f. = $(10 + 20 + f_1)$, i = 10, $f_2 = 40$

Substituting the values in the median formula

$$35 = 30 + \frac{85 - (10 + 20 + f_1)}{40} \times 10$$

$$35 = 30 + \frac{85 - 10 + 20 + f_1}{40} \times 10$$

$$35 = 30 + \frac{55 - f_1}{4}$$

$$30 + \frac{55 - f_1}{4} = 5 \quad \text{or } 55 - f_1 = 20 \quad \text{or} \quad f_1 = 35$$

Since $f_1 + f_2 = 60$, f_2 shall be 60–35 = 35. Thus the missing frequencies are $f_1 = 35$, $f_2 = 25$.

Calculation of Arithmetic Mean

Variable	*f*	*m.p.*	*(m – 35)/10* *d*	*fd*
0–10	10	5	– 3	– 30
10–20	20	15	– 2	– 40
20–30	35	25	– 1	– 35
30–40	40	35	0	0
40–50	25	45	+ 1	+ 25
50–60	25	55	+ 2	+ 50
60–70	15	65	+ 3	+ 45
	N = 170			Σfd = 185

$$\overline{X} = A + \frac{\Sigma fd}{N} \times i$$

A = 35, Σfd = 15, N = 170, i = 10

$$\overline{X} = 35 + \frac{15}{170} \times 10$$

$= 35 + .882 = 35.882$

Hence the arithmetic mean of the completed table is 35.882.

Example 19:

From the following data calculate median:

Marks Less than 5	***No. of Students***	***Marks Less than 5***	***No. of Students***
	29	*30*	*644*
" 10	*224*	*" 35*	*650*
" 15	*465*	*" 40*	*653*
" 20	*582*	*" 45*	*655*
" 25	*634*		

Solution:

Since we are given cumulative frequencies; first find simple frequencies and then calculate median.

Marks	***No. of Students*** f	***c.f.***
0–5	29	29
5–10	195	224
10–15	241	465
15–20	117	582
20–25	52	634
25–30	10	644
30–35	6	650
35–40	3	653
40–45	2	655

$$\text{Med.} = \text{Size of } \frac{N}{2}\text{th item} = \text{Size of } \frac{655}{2} = 327.5\text{th item}$$

Median lies in the class 10–15

$$\text{Median} = L + \frac{N/2 - c.f.}{f} \times i$$

$$L = 10,\ N/2 = 327.5,$$

$$c.f. = 224,\ f = 245,\ i = 10$$

$$\text{Median} = 10 + \frac{327.5 - 224}{241} \times 10 = 10 + 4.27 = 14.27.$$

Calculation of Median when Class Intervals are Unequal

When the class intervals are unequal, the frequencies need not be adjusted to make the class intervals equal and the same formula for interpolation can be applied as discussed above.

Example 20:

Calculate the lower and upper quartiles, third decile and 20th percent from the following data:

Variable:	*2.5*	*7.5*	*12.5*	*17.5*	*22.5*
Frequency:	*7*	*18*	*25*	*30*	*20*

Solution:

Since we are given mid-points, we will first find the lower and upper limits of the various classes. The method for finding these limits is to take the difference between the two central values, divide it by 2, deduct the values so obtained from the lower limit and add it to the upper limit, in the given cases $\frac{7.5 - 2.5}{2} = \frac{5}{2} = 2.5$. The first class shall be 0.5, second 5–10, etc.

Calculation of Q_1, Q_2, Q_3, P_{20}

Class group	*f*	*c.f.*
0–5	7	7
5–10	18	25
10–15	25	50
15–20	30	80
20–25	20	100
	N = 100	

Lower Quartile Q_1 = Size of $\frac{N}{4}$ th item = $\frac{100}{4}$ = 25th item

Q_1 lies in the class 5–10.

$$Q_1 = L + \frac{N/4 - c.f.}{f} \times i$$

$$L = 5,\ N/4 = 25,$$

$$c.f. = 7,\ f = 18,\ i = 5$$

$$Q_1 = 55 + \frac{25-7}{18} \times = 5 + 5 = 10$$

Upper Quartile Q_3 = Size of $\frac{3N}{4}$th item $= \frac{3 \times 100}{4}$ = 75th item

Q_3 lies in the class 15–20

$$Q_3 = L + \frac{3N/4 - c.f.}{f} \times i$$

$$L = 15,\ 3N/4 = 75,$$

$$c.f. = 50,\ f = 30,\ i = 5$$

$$\therefore \quad Q_3 = 15 + \frac{75-50}{30} \times 5 = 15 + 4.17 = 19.17$$

Third Quartile D_3 = Size of $\frac{3N}{10}$th item $= \frac{3 \times 100}{10}$ = 30th item

D_3 lies in the class 10–15

$$D_3 = L + \frac{3N/10 - c.f.}{f} \times i$$

$$L = 10,\ 3N/10 = 30,\ c.f. = 25,\ f = 25,\ i = 5$$

$$\therefore \quad D_3 = 10 + \frac{30-25}{25} \times 5 = 10 + 1 = 11$$

Twentieth Percentile P_{20} = Size of $\frac{20N}{100}$th item

$$= \frac{20 \times 100}{100} = 20\text{th item}$$

$$P_{20} = L + \frac{20N/100 - c.f.}{f} \times i$$

$$L = 5,\ 20N/100 = 20,\ c.f. = 7,\ f = 18,\ i = 5$$

$$\therefore \quad P_{20} = 5 + \frac{20-7}{18} \times 5 = 5 + 3.61 = 8.61.$$

DETERMINATION OF MEDIAN, QUARTILES, ETC., GRAPHICALLY

Median can be determined graphically by applying any of the following two methods:

(a) Draw two ogives—one by 'less than' method and the other by 'more than' method. From the point where both these curves intersect each other draw a perpendicular on the X-axis. The point where this perpendicular touches the X-axis gives the value of median.

(b) Draw only one ogive by 'less than' method. Take the variable on the X-axis and frequency on the Y-axis. Determine the median value by the formula: median = size of $\frac{N}{2}$th item. Locate this value on the Y-axis and from it draw a perpendicular on the cumulative frequency curve. From the point where it meets the ogive draw another perpendicular on the X-axis and the point where it meets the X-axis is the median.

The other partition values like quartiles, deciles, etc., can also be determined graphically by following method No. 2.

Example 21:

Calculate median from the following data:

Marks	*0–10*	*10–30*	*30–60*	*60–80*	*80–90*
No. of Students	*5*	*15*	*30*	*8*	*2*

Solution:

Since class intervals are unequal, let us first convert it to a distribution with equal class intervals on the assumption that the frequencies are equally distributed throughout a class.

Calculation of Median

Marks	*f*	*c.f.*	*Marks*	*f*	*c.f.*
0–10	5	5	60–80	8	58
10–30	15	20	80–90	2	60
30–60	30	50			

$$\text{Med.} = \text{Size of } \frac{N}{2}\text{th item} = \text{Size of } \frac{60}{2} = 30\text{th item}$$

Median lies in the class 30–60

$$\text{Median} = L + \frac{N/2 - c.f.}{f} \times i$$

$$L = 300,\ N/2 = 30,\ c.f. = 20,\ f = 30,\ i = 30$$

$$\text{Median} = 30 + \frac{30 - 20}{30} \times 30 = 30 + 10 = 40.$$

Mathematical Property of Median

The sum of the deviations of the items from median, ignoring sign, is the least. For example, the median of 4, 6, 8, 10, 12 is 8. The deviations from 8 ignoring signs are 4, 2, 0, 2, 4 and the total is 12. This total is smaller than the one obtained if deviations are taken from any other value. Thus if deviations are taken from 7, values ignoring signs would be 3, 1, 1, 3, 5, and the total 13.

Merits and Limitations of Median

Merits

(a) Perhaps the greatest advantage of median is, however, the fact that the median actually does indicate what many people incorrectly believe the arithmetic mean indicates. The median indicates the value of the middle item in the distribution. This is a clear-cut meaning and makes the median a measure that can be easily explained.

(b) The value of median can be determined graphically whereas the value of mean cannot be graphically ascertained.

(c) It is the most appropriate average in dealing with qualitative data, *i.e.*, where ranks are given or there are other types of items that are not counted or measured but are scored.

(d) In markedly skewed distributions such as income distributions or price distributions where the arithmetic mean would be distorted by extreme values, the median is especially useful. Consequently, the median income for some purposes be regarded as a more representative figure, for half the income earners must be receiving at least the median income. One can say as many receive the median income and as many do not.

(e) It is especially useful in case of open-end classes since only the position and not the values of items must be known. The median is also recommended if the distribution has unequal classes, since it is easier to compute than the mean.

(f) Extreme values do not affect the median as strongly as they do the mean. For example, the median of 10, 20, 30, 40 and 150 would be 30 whereas the mean 50. Hence very often when extreme values are present in a set of observations, the median is a more satisfactory measure of the central tendency than the mean.

Limitations

(a) For calculating median it is necessary to arrange the data; other averages do not need any arrangement.

(b) The value of median is affected more by sampling fluctuations than the value of the arithmetic mean.

(c) Since it is a positional average, its value is not determined by each and every observation.

(d) It is not capable of algebraic treatment. For example, median cannot be used for determining the combined median of two or more groups as is possible in case of mean. Similarly, the median wage of a skewed distribution times the number of workers will not give the total payroll. Because of this limitation the median is much less popular as compared to the arithmetic mean.

(e) It is erratic if the number of items is small.

(f) The median, in some cases, cannot be computed exactly as the mean. When the number of items included in a series of data is even, the median is determined approximately as the mid-point of the two middle items.

Usefulness

The median is useful for distributions containing open-end intervals since these intervals do not enter its computation. Also since the median is affected by the number rather than the size of items, it is frequently used instead of the mean as a measure of central tendency in cases where such values are likely to distort the mean.

RELATED POSITIONAL MEASURE

Besides median, there are other Measure which divide a series into equal parts. Important amongst these are quartiles, deciles and percentiles. Quartiles are those values of the variate which divide the total frequency into four equal parts, deciles divide the total frequency into 10 equal parts and the percentiles divide the total frequency into 100 equal parts. Just as one point divides a series into two parts, three points would divide it into four parts, 9 points into 10 parts and 99 points into 100 parts. Consequently, there re only 3 quartiles, 9 deciles and 99 percentiles for a series. The quartiles are denoted by symbol Q, deciles by D and percentiles by P. The subscripts 1, 2, 3, etc., beneath Q, D, etc., would refer to the particular value that we want to compute. Thus Q_1 would denote first quartile, Q_2 second quartile, Q_3 third quartile, D_1 first decile, D_8 8th decile, P1 first percentile and P_{60} 60th percentile, etc.

Graphically, any set of these partition values divides the area of the frequency curve or histogram into equal parts. If vertical lines are drawn as third quartiles, for example, the area of the histogram will be divided by these lines into four equal parts. The 9 deciles divide the area of the histogram or frequency curve into 10 equal parts and the 99 percentiles divide the area into 100 equal parts.

In economics and business statistics quartiles are more widely used than deciles and percentiles. The quartiles are the points on the X-scale that divide the distribution into four equal parts. Obviously, there are three quartiles, the second coinciding with the median. More precisely stated, the lower quartile Q_1 is that point on the X-scale such that one-fourth of the total frequency is less than Q_1 and three-fourths is greater than Q_1. The upper quartile Q_3, is that point on the X-scale such that three-fourths of the total frequency is below Q_3 and one-fourth is above it.

The deciles and percentiles are important in psychological and educational statistics concerning grades, rates, ranks, etc. they are of use in economics and business statistics in personnel work, productivity ratings and other such situations.

It should be noted that quartiles, deciles, etc., are not averages. They are Measure of dispersion and as such shall be discussed in detail in the next chapter. Here only a passing reference is made. The method of computing these partition values is the same as discussed for median.

Just as quartiles divide the series into 4 equal parts, pantiles divide into 5 equal parts, septiles into 7 equal parts an doctiles into 8 equal parts. However, these partition values are rarely used in practice.

Computation of Quartiles, Percentiles, etc.

The procedure for computing quartiles deciles, etc., is the same as the median. While computing these values in individual and discrete series we add 1 to N whereas in continuous series we do not add 1.

Thus $Q_1 = \text{Size of } \frac{N+1}{4} \text{ item}$

(individual observations and discrete series)

$$Q_1 = \text{Size of } \frac{N}{4} \text{ (in continuous series)}$$

$$Q_3 = \text{Size of } \frac{3(N+1)}{4}\text{th item (in individual and discrete series)}$$

$$Q_3 = \text{Size of } \frac{3N}{4} \text{th item (in continuous series)}$$

$$D_4 = \text{Size of } \frac{4(N+1)}{10} \text{th item (in individual and discrete series)}$$

$$Q_3 = \text{Size of } \frac{4N}{10} \text{th item (in continuous series)}$$

$$P_{60} = \text{Size of } \frac{60(N+1)}{100} \text{th item (in individual and discrete series)}$$

$$Q_3 = \text{Size of } \frac{60N}{100} \text{th item (in continuous series)}$$

Example 22:

Draw an ogive for the following distribution. How many workers earned wages between Rs. 1365 and Rs. 1430? Also calculate the median wage.

Wages	***No. of workers (Rs.)***	***Wages***	***No. of workers (Rs.)***
1000–1100	*6*	*1400–1500*	*16*
1100–1200	*10*	*1500–1600*	*14*
1200–1300	*22*	*1600–1700*	*12*
1300–1400	*30*		

Solution:

Wages less than (Rs.)	***c.f.***	***Wages less than (Rs.)***	***c.f.***
1100	6	1500	84
1200	16	1600	98
1300	38	1700	110
1400	68		

Number of workers whose wages are between Rs. 1365 and Rs. 1430 is 16 as shown below

Workers Getting Wages Between

Rs. 1365 and 1430

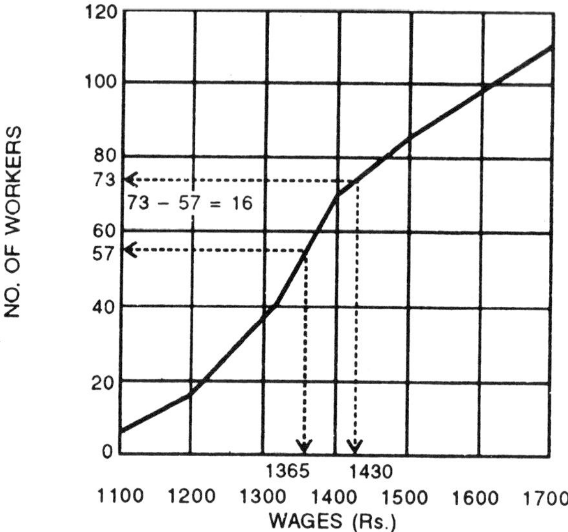

Number of workers whose wages are less than Rs. 1430 = 73

Number of workers whose wages are less than Rs. 1365 = 57

Number of workers whose wages are between

1365 and 1430 = 73 – 57 = 16

Calculation of Median

Wages (Rs.)	*No. of workers*	*c.f.*	*Wages (Rs.)*	*No. of workers*	*c.f.*
1000–1100	6	6	1400–1500	16	84
1100–1200	10	16	1500–1600	14	98
1200–1300	22	38	1600–1700	12	110
1300–1400	30	68			

$$\text{Med.} = \text{Size of } \frac{N}{4} \text{ item} = \text{Size of } \frac{110}{2} = 55\text{th item}$$

Hence median lies in the class 1300–1400.

$$\text{Median} = L + \frac{N/2 - c.f.}{f} \times i$$

L = 1300, N/2 = 55,

c.f. = 38, f = 30, i = 100

$$\text{Median} = 1300 + \frac{55 - 38}{30} \times 100 = 1300 + 56.67 = \text{Rs. } 1356.67$$

Example 23:

Calculate the mode from the following data of the marks obtained by 10 students:

Sl. No.	*Marks obtained*	*Sl. No.*	*Marks obtained*
1	*10*	*6*	*27*
2	*27*	*7*	*20*
3	*24*	*8*	*18*
4	*12*	*9*	*15*
5	*27*	*10*	*30*

Solution:

Calculation of Mode

Size of item	*Number of time it occurs*	*Size of item*	*Number of time it occurs*
10	1	20	1
12	1	24	1
15	1	27	3
18	1	30	1
			Total 10

Since the item 27 occurs the maximum number of times, *i.e.*, 3, hence the modal marks are 27.

Note: Thus the process of determining mode in case of individual observations essentially involves grouping of data.

When there are two or more values having the same maximum frequency, one cannot say which is the modal value of hence mode is said to be ill-defined. Such a series is also shown as bimodal or multimodal. For example, observe the following data:

Income (in Rs.) 110 120 130 120 110 140 130 120 130 140

Size of items	110	120	130	140
No. of times it occurs	2	3	3	2

Since 120 and 130 have the same maximum frequency, *i.e.*, 3, mode is ill defined in the case.

Calculation of Mode—Discrete Series

In discrete series quite often mode can be determined just by inspection, *i.e.*, by looking to that value of the variable around which the items are most heavily concentrated. For example, observe the following data:

Size of garment	28	29	30	31	32	33
No. of persons wearing:	10	20	40	65	50	15

From the above data we can clearly say that the modal size is 31 because the value 31 has occurred the maximum number of times, *i.e.*, 65. However, where the mode is determined just by inspection, an error of maximum frequency and the frequency preceding it or succeeding it is very small and the items are heavily concentrated on either side. In such cases it is desirable to prepare a grouping table and an analysis table. These tables help us in ascertaining the modal class.

A grouping table has six columns. In column 1 the maximum frequency is marked or put in a circle; in column 2 frequencies are grouped in two's; in column 3 leave the first frequency and then group the remaining in two's; in column 4 group the frequencies in three's; in column 5 leave the first frequency and group the frequencies in three's; and in column 6 leave the first two frequencies and then group the remaining in three's. In each of these cases take the maximum total and mark it in a circle or by bold type.

After preparing the grouping table, prepare an analysis table. While preparing the table put column number on the left-hand side and the various probable values of mode on the right-hand side. The values against which frequencies are the highest are marked in the grouping table and then entered by means of a bar in the relevant 'box' corresponding to the values they represent.

The procedure of preparing grouping table and analysis table shall be clear from the following example:

Example 24:

Find the value of mode from the data given below:

Weight (kg)	*No. students*	*Weight (kg)*	*No. of students*
93–97	*2*	*113–117*	*14*
98–102	*5*	*118–122*	*6*
103–107	*12*	*123–127*	*3*
108–112	*17*	*128–132*	*1*

Solution:

By inspection mode lies in the class 108–112. But the real limits of this class are 107.5–112.5.

$$\text{Mode} = L + \frac{\Delta_1}{\Delta_1 + \Delta_2} \times i$$

$$L = 107.5,\ \Delta_1 = f_1 - f_0 = (17 - 12) = 5,$$

$$\Delta_2 = f_1 - f_2 = (17 - 14) = 3,\ i = 5$$

$$\text{Mode} = 107.5 + \frac{5}{5+3} \times 5 = 107.5 + 3.125 = 110.625$$

Thus the modal weight is 110.625 kg.

Example 25:

Calculate mode from the following data:

Marks	*No. of students*	*Marks*	*No. of students*
Above 0	*80*	*Above 60*	*25*
Above 10	*77*	*Above 70*	*16*
Above 20	*72*	*Above 80*	*10*
Above 30	*65*	*Above 90*	*8*
Above 40	*55*	*Above 100*	*0*
Above 50	*43*		

Solution:

Since this is cumulative frequency distribution, we first convert it into a simple frequency distribution.

Marks	*No. of students*	*Marks*	*No. os students*
0–10	3	50–60	15
10–20	5	60–70	12
20–30	7	70–80	6
30–40	10	80–90	2
40–50	12	90–100	8

By inspection the modal class is 50–60.

$M_o = L + \frac{\Delta_1}{\Delta_1 + \Delta_2} \times i$; $L = 50$; $\Delta_1 = (15 - 12) = 3$,

$\Delta_2 = (15 - 12) = 3$; $i = 10$;; $M_o = 50 + \frac{3}{3 + 30} \times 10 = 50 + 5 = 55.$

Example 26:

Calculate the value of mode for the following data:

Marks :	*10*	*15*	*20*	*25*	*30*	*35*	*40*
Numbers :	*8*	*12*	*36*	*25*	*28*	*18*	*9*

Solution:

Since it is difficult to say by inspection as to which is the modal value, we prepare grouping and analysis tables.

Calculation of Mode

x	f	II	III	IV	V	VI
10	8	20		56		
15	12		48		83	
20	36	71				99
25	35		63	81		
30	28	46			55	
35	18		27			
40	9					

Analysis Table

Col. No.	20	25	30
I	1		
II	1	1	
III		1	1
IV		1	1
V	1	1	
VI	1	1	1
	4	5	3

Corresponding to the maximum total 5, the value of the variable is 25. Hence modal value is 25.

CALCULATION OF MODE—CONTINUOUS SERIES

Steps:

(i) By preparing grouping table and analysis table or by inspection ascertain the modal class.

(ii) Determine the value of mode by applying the following formula:

$$Mo = L + \frac{\Delta_1}{\Delta_1 + \Delta_2} \times i$$

where, L = lower limit of the modal class; Δ_1 = the difference between the frequency of the modal class and the frequency of the pre-modal class, *i.e.*, preceding class (ignoring signs); Δ_2 = the difference between the frequency of the modal class and the frequency of the post-modal class, *i.e.*, succeeding class (ignoring signs); i = the class interval of the modal class.

Another form of this formula is

$$Mo = L + \frac{f_1 - f_0}{2f_1 - f_0 - f_2} \times i$$

where, L = lower limit of the modal class; f_1 = frequency of the modal class; f_0 = frequency of the class preceding the modal class; f_2 = frequency of the class succeeding the modal class.

There may be two values which occur with equal frequency. The distribution is then called bimodal. The following is a graph of bimodal distribution:

In a bimodal distribution the value of mode cannot be determined with the help of formula given above. If plotted data produce a bimodal distribution, the data themselves should be questioned. Quite often such a condition is caused when the side of the sample is small; the difficulty can be remedied by increasing the sample size. Another common cause is the use of non-homogeneous data. In instances where a distribution is bimodal and nothing can be done to change it, the mode should not be used as a measure off central tendency.

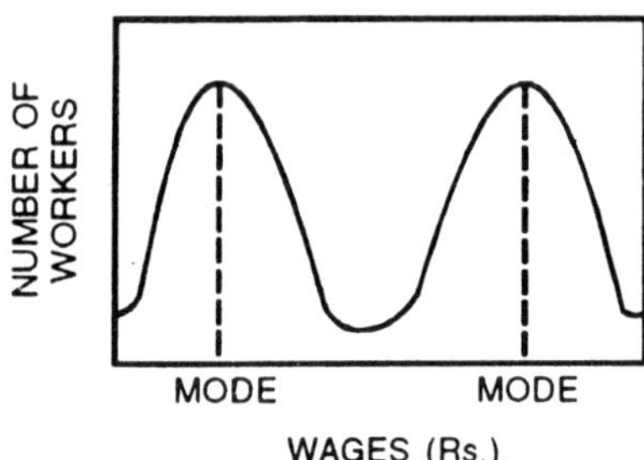

Where mode is ill-defined its value may be ascertained by the following formula based upon the relationship between mean, median and mode:

$$\text{Mode} = 3 \text{ Median} - 2 \text{ Mean}$$

This measure is called the emperical mode.

Mode

The mode or the modal value is that value in a series of observations which occurs with the greatest frequency. For example, the mode of the series 3, 5, 8, 5, 4, 5, 9, 3 would be 5, since this value occurs more frequently than any of the others.

The mode is often said to be that value which occurs most often in the data, that is, with the highest frequency. While this statement is quite helpful in interpreting the mode, it cannot safely be applied to any distribution, because of the vagaries of sampling. Even fairly large samples drawn from a statistical population with a single well defined mode may exhibit very erratic fluctuations in this average if the mode is defined as that exact value in the ungrouped data of each sample which occurs most frequently. Rather it should be thought as the value about which the items are most closely concentrated. It is the value which has the greatest frequency density concentrated. It is the value which has the greatest frequency density in the immediate neighbourhood. For this reason mode is also called the most typical or fashionable value of a distribution.

The following diagram shows the modal value:

The value of the variable at which the curve reaches a maximum is called the mode. It is the value around which the items tend to be most heavily concentrated.

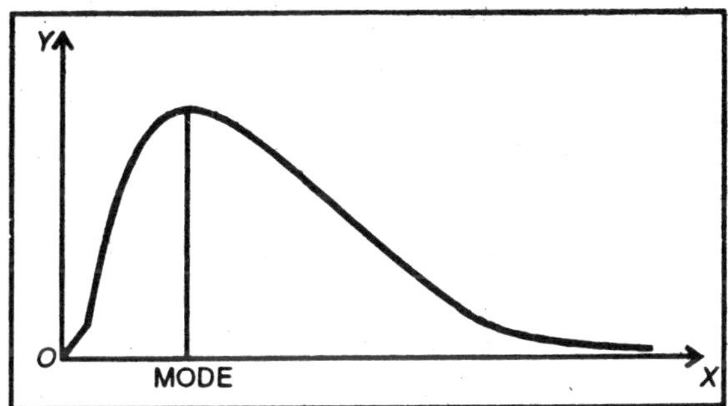

Although mode is that value which occurs most frequently yet it does not follow that its frequency represents a majority out of all the total number of frequencies. For example, in the election of college president the votes obtained by three candidates contesting for presidentship out of a total of 816 votes polled are as follows:

Mr. X 268; Mr. Y 278; Mr. Z 270 : Total 816

Mr. Y will be elected as president because he has obtained highest votes. But it will be wrong to say that he represents majority because there are more votes against him (268 + 270 = 538) than those for him.

There are many situations in which arithmetic mean and median fail to reveal the true characteristic of data. For example, when we talk of most common wage, most common income, most common height, most common size of shoe or ready-made garments, we have in mind mode and not the arithmetic mean or median discussed earlier. The mean does not always provide an accurate reflection of the data due to the presence of extreme items. Median may also prove to be quite unrepresentative of the data owing to an uneven distribution of the series. For example, the values in the lower half of a distribution range from, say, Rs. 10 to Rs. 100, while the same number of items in the upper half of the series range from Rs. 100 to Rs. 6,000 with most of them near the higher limit. In such a distribution the median value of Rs. 100 will provide little indication of the true nature of the data.

Both these shortcomings may be overcome by the use of mode which refers to the value which occurs most frequently in a distribution. Moreover, mode is the easiest to compute since it is the value corresponding to the highest frequency. For example, if the data are:

Size of shoes	5	6	7	8	9	10	11
No. of persons	10	20	25	40	22	15	6

The modal size is '8' since it appears maximum number of times in the series.

Calculation of Mode

Determining the precise value of the mode of a frequency distribution is by no means an elementary calculation. Essentially it involves fitting mathematically some appropriate type of frequency curve to the grouped data and determination of the value on the X-axis below the peak of the curve. However, there are several elementary methods of estimating the mode. These methods have been discussed for individual observation, discrete series and continuous series.

Calculation of Mode—Individual Observations

For determining mode count the number of times the various values repeat themselves and the value occurring maximum number of times is the modal value. The more often the modal value appears relatively, the more valuable the measure is an average to represent data. The use of non-homogeneous data. In instances where a distribution is bimodal and nothing can be done to change it. The mode should not be used as a measure of central tendency.

Where mode is ill-defined, its value may be ascertained by the following formula based upon the relationship between mean, median and mode:

$$\text{Mode} = 3\text{ Median} - 2\text{ Mean}$$

This measure is called the empirical mode.

Example 27:

Draw a histogram for the following distribution and find the modal wage and check the value by direct calculation.

Wages (in Rs.) :	*10–15*	*15–20*	*20–25*	*25–30*	*30–35*	*35–40*	*40–45*
No. of Workers :	*60*	*140*	*110*	*150*	*120*	*100*	*90*

Solution:

The histogram of this data is given below:

It is clear from the histogram that the modal value is:

Direct Calculation

Mode lies in the class 25–30.

$$M_0 = L + \frac{\Delta_1}{\Delta_1 + \Delta_2} \times i$$

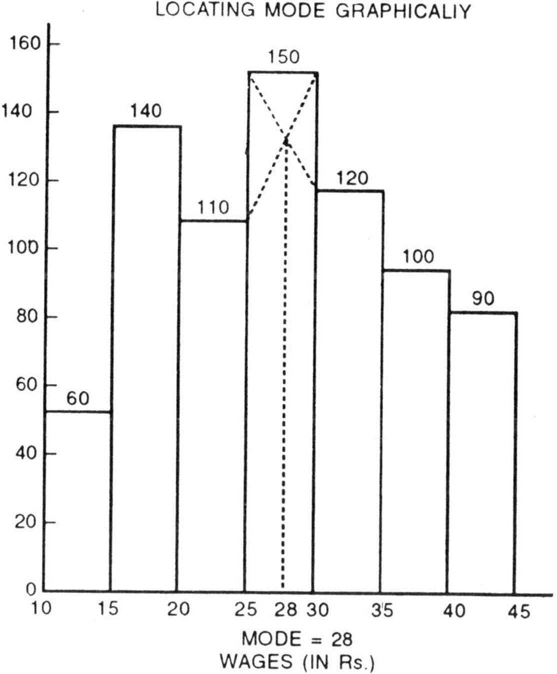

$$L = 25,\ \Delta_1 = (150 - 110) = 40,$$

$$\Delta_2 = (150 - 120) = 30,\ i = 10$$

$$M_0 = 25 + \frac{40}{40 + 30} \times 5$$

$$= 25 + 2.86 = 27.86.$$

The slight difference in the two answers is due to the difficulty of reading very precisely from the graph.

Mode can also be determined from a frequency polygon in which case a perpendicular is drawn on the base from the apex of the polygon and the point where it meets the base gives the modal value.

However, graphic method of determining mode can be used only where there is one class containing the highest frequency. If two or more classes have the same highest frequency, mode cannot be determined graphically. For example, for the data given below mode cannot be graphically ascertained.

Size of shoes	*No. of persons wearing*	*Size of shoes*	*No. of persons wearing*
2–4	10	8–10	8
4–6	15	10–12	2
6–8	15		

Merits and Limitations of Mode

Merit

The main merits of mode are:

(a) It can be used to describe qualitative phenomenon. For example, if we want to compare the consumer preferences for different types of products, say, soap, toothpaste, etc., or different media of advertising we should complete the modal preferences expressed by different groups of people.

(b) The value of mode can also be determined graphically whereas the value of mean cannot be graphically ascertained.

(c) By definition mode is the most typical or representative value of a distribution. Hence, when we talk of modal wage, modal size of shoe or modal size of family it is this average that we refer to. The mode is a measure which actually does indicate what many people incorrectly believe the arithmetic mean indicates. The mode is the most frequently occurring value. If the modal wage in a factory is Rs. 916 then more workers receive Rs. 916 than any other wage. This is what many believe the "average" wage always indicates, but actually such a meaning is indicated only if the average used is the mode.

(d) Like median, the mode is not unduly affected by extreme values. Even if the high values are very high and the low values are very low we choose the most frequent value of the data to the modl value: for example, the mode of 10, 2, 5, 10, 5, 60, 5, 10, 60 is 10 as this value, *i.e.*, 10 has occurred most often in the data set.

(e) Its value can be determined in open-end distributions without ascertaining the class limits.

Limitations

The important limitations of this average are:

(a) It is not a rigidly defined measure. There are several formulae for calculating the mode, all of which usually give somewhat different answers in fact, mode is the mot unstable average and its value is difficult to determine.

(b) While dealing with quantitative data, the disadvantages of the mode outweigh its good features and hence it is seldom used.

(c) The value of mode cannot always be determined. In some cases we may have a bimodal series.

(d) It is not capable of algebraic manipulations. For example, from the modes of two sets of data we cannot calculate the overall mode of the combined data. Similarly, the modal wage times the number of workers will not give the total payroll—except, of course, when the distribution is normal and then the mean, median and mode are all equal.

(e) The value of mode is not based on each and every item of the series.

Usefulness: The mode is employed when the most typical value of a distribution is desired. It is the most meaningful measure of central tendency in case of highly skewed or non-normal distributions, as it provides the best indication of the point of maximum concentration.

Relationship Among Mean, Median and Mode

A distribution in which the values of mean, median and mode coincide (*i.e.*, mean = median = mode) is known as a symmetrical distribution. Conversely stated, when the values of mean, median and mode are not equal to distribution is known as asymmetrical or skewed. In moderately skewed or asymmetrical distributions a very important relationship exists among mean, median and mode. In such distributions the distance between the mean and the median is about one-third the distance between the mean and the mode as will be clear from the diagram given below.

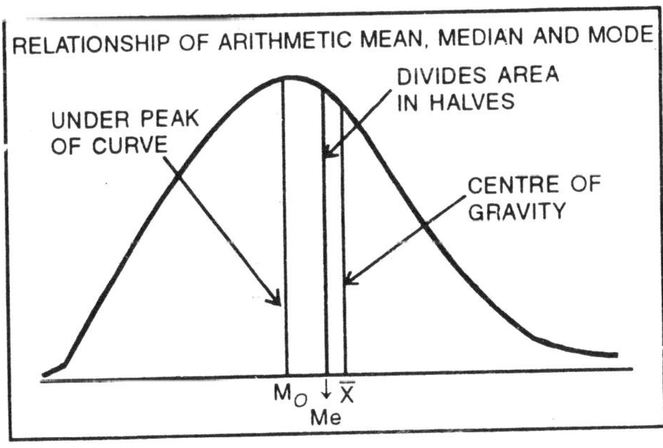

Karl Pearson has expressed this relationship as follows:

Mode = Mean – 3 [Mean –Median]

Mode = 3 Median – 2 Mean

and Median = Mode + $\frac{2}{3}$ [Mean – Mode]

If we know any of the two values out of the three, we can compute the third from these relationships. The following example will illustrate this point:

Example 28:

Daily income of ten families of a particular place is given below.

85 70 15 75 500 8 45 250 40 36

Solution:

Calculation of Geometric Mean

X	*Log X*	*X*	*Log X*
85	1.9294	8	0.9031
70	1.8451	45	1.6532
15	1.1761	250	2.3979
75	1.8751	40	1.6021
500	2.6990	36	1.5563

$$\text{G.M.} = \text{AL}\left(\frac{\sum \log x}{N}\right) = \text{AL}\frac{17.6373}{10} = \text{AL } 1.7637 = 58.03.$$

Example 29:

From the following data of weight of 122 persons determine the modal weight;

Solution:

By inspection it is difficult the modal weight:

Weight (in ibs.)	*No. of persons*	*Weight (in ibs)*	*No. of persons*
100-110	4	140-150	33
110-120	6	150-160	17
120-130	20	160-170	8
130-140	32	170-180	2

By inspection it is difficult to say which is the modal class. Hence we prepare a grouping table and an analysis table.

Grouping Table

Weight (in Ibs.)	No. of persons I	II	III	IV	V	VI
100-110	4					
		10				
110-120	6			30		
			26			
120-130	20				58	
		52				
130-140	32					85
			65			
140-150	33			82		
		50			58	
150-160	17					27
			25			
160-170	8					
		10				
170-180	2					

Class in which mode is expected to lie

Col. No.	*120-130*	*130-140*	*140-150*
1			1
2	1	1	
3		1	1
4		1	1
5	1	1	1
6	1	1	1
	Total 3	5	5

This is a bi-modal series. Hence mode has to be determined by plying the formula:

Mode = 3 Median – 2 Mean

Weight in ib.	*m*	*No. of persons* *f*	*c.f.*	*(m-135)/10* *d*	*fd*
100-110	105	4	4	–3	–12
110-120	115	6	10	–2	–12
120-130	125	20	30	–1	–20
130-140	135	32	62	0	0
140-150	145	33	95	+1	+33
150-160	155	17	112	+2	–34
160-170	165	8	122	+3	–24
170-180	175	2	122	+4	+8
		N = 122			Σfd = 55

$$\overline{X} = A + \frac{\sum fd}{N} xi$$

$A = 135, \Sigma fd = 55, N = 122, i = 10$

$$\overline{X} = 135 + \frac{55}{122} \times 135 + 4.51 = 139.51$$

$$\text{Med.} = \text{size of } \frac{N}{2}\text{th item} = \text{size of } \frac{122}{2} = 61\text{st item.}$$

Hence median lies in the class 130–140.

$$\text{Median} = L + \frac{N/2 - c.f.}{2f} \times i$$

$L = 130, N/2 = 61, c.f. = 30, f = 32, i = 10$

$$\text{Median} = 130 + \frac{(61 - 30)}{32} \times 10 = 130 + \frac{310}{32} = 139.69$$

Mode = 3 Median = 2 Mean

Mode = (3 × 139.69) – (2 × 139.51) = 419.07 – 279.02 = 140.05.

Hence model weigh is

Mode when Class Intervals are Unequal

The formula for calculating the value of mode given above is applicable only where there are equal class intervals. If the class intervals are unequal, then we must make them equal before we start computing the value of mode.

The class interval should be made equal and frequencies adjusted on the assumption that they are equally distributed thought the class.

Locating Mode Graphically

In a frequency distribution the value of mode can also be determined graphically. The steps in calculation are:

(a) Draw a histrogram of the given data.

(b) Draw two lines diagonally in the inside of the modal class bar, starting from each upper corner of the bar to the upper corner of the adjacent bar.

(c) Draw a perpendicular line from the intersection of the two diagonal lines to the X-axis (horizontal scale) which gives us the modal value.

Example 30:

Calculate geometric mean from the following data:

125 1462 38 7 0.22 0.08 12.75 0.5

Solution:

Calculation of G.M.

X	***Log X***
125	2,0969
1462	3,1650
38	2,5798
7	0.8451
0.22	.3424
0.08	.9031
12.75	1.1055
0.5	.6990
	Σ log X = 6.7360

$$\text{G.M.} = \text{AL}\left(\frac{\sum \log X}{N}\right) = \text{AL}\left(\frac{6.7360}{8}\right)$$

$$= \text{Al }(0.8421 = 6.952).$$

Calculation of Geometric Mean – Discrete Series

$$\text{G.M.} = \text{Antilog}\left(\frac{\sum f \log X}{N}\right)$$

Steps : Find the logarithms of the variable X.

(i) Multiply these logarithms with the respective frequencies and obtain the total $\sum f \log X$.

(ii) Divide $\sum f \log X$ by the total frequency and take the antilog of the value so obtained.

Calculation of Geometric Mean–Continuous Series

$$\text{G.M.} = \text{Antilog}\left(\frac{\sum f \log X}{N}\right)$$

(i) Find out the mid-points of teh classes and take their logarithms.

(ii) Multiply these logarithms with the respective frequencies of each class and obtain the total $\sum f \log m$.

(iii) Divide the total obtained in step (ii) by the total frequency and take the antilog of the value so obtained.

Example 31:

(i) In a moderately asymmetrical distribution, the mode and mean are 32.1 and 35.4 respectively. Find out the value of Median.

Solution:

(i) Mode = 3 Median – 2 Mean

Given mean = 35.4, mode = 32.1

32.1 = 3 median – 2 × 35.4

3 median = 32.1 + 70.8 = 102.9

or median = 102.9/3 = 34.3

Given median = 20.6, mode = 26, find mean.

(ii) Mode = 3 Median – 2 Mean

Given mean = 20.6, mode = 26

26 = 3 × 20.6 – 2 mean

26 = 61.8 – 2 mean.

2 mean = 61.8 – 26 = 35.8 or mean = 17.9

GEOMETRIC MEAN

Geometric mean is defined as the Nth root of the product of N items or values. If there are two items, we take the square root: if there are three items, the cube root; and so on. Symbolically.

$$G.M. = \sqrt{(X_1) \times (X_2) \times (X_3) \times \ldots\ldots\ldots X_n}$$

here X_1, X_2, X_3, etc. refer to the various items of the series.

Thus the geometric mean of 3 values 2, 3 4, would be:

$$G.M. = \sqrt[3]{2 \times 3 \times 4} = \sqrt{24} = 2.885$$

When the number of items is three or more the task of multiplying the numbers and of extracting the root becomes excessively difficult. To amplify calculations logarithms are used. Geometric mean then is calculated as follows:

$$\log G.M. = \frac{\log X_1 + \log X_2 + \ldots + \log X_n}{N}$$

$$\text{or } \log G.M. = \left(\frac{\sum \log X}{N}\right) \quad \therefore G.M. = \text{Anti}\log\left(\frac{\sum \log X}{N}\right)$$

$$\text{In discrete series G.M.} = \text{Antilog}\left(\frac{\sum f \log X}{N}\right)$$

$$\text{In continuous series G.M.} = \text{Antilog}\left(\frac{\sum f \log m}{N}\right)$$

Properties of Geometric Mean

The following are two important mathematical properties of geometric mean:

(a) The product of the value of series will remain unchanged when the value of geometric mean is substituted for each individual value. For example, the geometric mean for series 2, 4, 8 is 4 : therefore, we have

$$2 \times 4/ \times 8 = 64 = 4 \times 4 \times 4$$

(b) The sum of the deviations of the logarithms of the original obwervations above or below the logarithm of the geometric mean is equal. this also means that the value of the geometric mean is such as to balance the ratio deviations of the observations from it. Thus, using the same pervious numbers, we find that

$$\left(\frac{4}{2}\right)\left(\frac{4}{4}\right) = 2 = \left(\frac{8}{4}\right)$$

Because of this property this measure of central value is especially adapted to average ratios, rates of change, and logarithmically distributed series.

Calculations of Geometric Mean—Individual Observations

$$\text{G.M.} = \text{Antilog}\left(\frac{\sum \log X}{N}\right)$$

(a) Take the logarithms of the variable X and obtain the total Σ log X.

(b) Divides S log X by n and take the antilog of the value so obtained. This gives the value of geometric mean.

Σ log X

If $f_1, f_2, f_3, \ldots f_n$—represents frequencies of

$X_1, X_2, X_3, \ldots X_n$ respectively then

$$\text{G.M.} = \sqrt[3]{\frac{(X_1 . X_1 . X_1 \ldots . f_2 \text{ times}) (X_2 . X_2 . X_3 \ldots . f_2 \text{ times})}{(X_2 . X_3 . X_4 \ldots . f_2 \text{ times}) (X_n . X_n . X_n \ldots . \text{ times})}}$$

$$X_1^{f_1} . X_2^{f_2} . X_n^{f_n}$$

Taking logarithms of both the sides to simplify calculation

$$\log \text{G. M.} = \frac{\log (X_1\ f_1 \times X_2\ f_2 \times X_3 \ldots . X_n\ f_n}{N} \qquad [\because N = \Sigma f]$$

$$= \frac{\log X_1^{f_2} \log X_2^{f^2} \log X_3^{f^3} \ldots . \log X_{f_n}}{N}$$

$$= \frac{f_1 \log X_1 + f_2 \log X_2 + f_2 \log X_2 \ldots . f^n \log X_n}{N}$$

$$= \frac{\sum f \log X}{N}$$

$$\text{G.M.} = \text{Antilog}\left(\frac{\sum \log X}{N}\right)$$

Example 32:

Find the Geometric mean for the data given below:

Marks	*Frequency*	*Marks*	*Frequency*
4-8	*6*	*24-28*	*12*
8-12	*10*	*28-32*	*10*
12-16	*18*	*32-36*	*6*
16-20	*30*	*36-40*	*2*
20-24	*15*		

Solution:

Calculation of Geometric Mean

Marks	*m.p.m*	*f*	*log m*	*f × log m*
4-8	6	6	0.7782	4.6692
8-12	10	10	1.0000	10.0000
12-16	14	18	1.1461	20.6298
16-20	18	30	1.2553	37.6590
20-24	22	15	1.3424	20.1360
24-28	26	12	1.4150	16.9800
28-32	30	10	1.4771	14.7710
32-36	34	6	1.5315	9.1890
36-40	38	2	1.5798	3.1596
		N = 109	Σf × log m = 137.1936	

$$\text{G.M.} = \text{AL}\left(\frac{\sum f \log X}{N}\right) = \text{AL}\left(\frac{137.1936}{109}\right) = \text{A.L. } 1.2587 = 18.14$$

Uses of Geometric Mean

Geometric mean is specially useful in teh following cases:

(a) The geometric mean is used to find the average per cent increase in sales, production, population or other economic or business series. For example, from 1996 to 1998 prices increased by 5, 10 and 18 per cent respectively. The average annual increase is not 11 per cent $\left(\frac{5+10+18}{3} = 11\right)$ as given by the arithmetic average but 10.9 per cent as obtained by the geometric mean.

(b) Geometric mean is theoretically considered to be the best average in the construction of index numbers. It satisfies the time reversal test and gives equal weight to equal ratio of change.

(c) This average is most suitable when large weights have to be given to small items and small weights to large items situations which we usually come across in social and economic fields.

The following examples illustrate the use of geometric mean:

Example 33:

Find the average rate of increase in population which in the first decade has increased by 20%, in the second decade by 30% and in the third decade by 40%.

Solution:

Calculation of Geometric Mean

Decade	*% Rise*	*X Population at the end of the decade taking population of the previous decade as 100*	*log X*
1st	*20*	*120*	*2.0792*
2nd	*30*	*130*	*2.1139*
3rd	*40*	*140*	*2.1461*
			Σ *log X = 6.3292*

$$G.M = A.L. \left(\frac{\sum \log X}{N}\right) = AL\left(\frac{6.3392}{3}\right) = A.L\ (2.1131) = 129.7$$

Thus, the average rate of increase in population is (129.7 – 100) 29.7 per cent predicate:

Example 34:

The population of a country has increased from 84 million in 1988 to 108 million in 1998. Find the annual rate of growth of population.

Solution:

Let r be the rate of growth. Applying the compound interest formula:

$$P_n = P_0(1 + r)^n$$

$$84(1 + r)^{10} = 108$$

Taking

$$\text{logarithms,} \quad \log(1 + r) = \frac{\log 108 - \log 84}{10} = \frac{2.0334 - 1.9243}{10}$$

$$= 0.0109$$

$$1 + r = 1.026;\ r = 0.026 = 2.6\%.$$

Example 35:

If arithmetic mean and geometric mean of two values are 10 and 8 respectively, find values.

Solution:

It two values are a and b then $\frac{a + b}{2} = 10\ \sqrt{ab} = 8$ or $ab = 64$

$$a + b = 20,\ ab = 64$$

$$(a - b) = \sqrt{(a + b)^2 - 4ab} = \pm 12$$

$$a + b = 20,\ a - b = \pm 12$$

Hence a = 16 and b = 4. The values are 16, 4.

Example 36:

The price of a commodity increased by 5% from 1995 to 1996, 8% from 1996 to 1997 and 77% from 1997 to 1998. The average increase from 1996 to 1998 is quoted as 26% and not 30%. Explain and verify the result.

Solution:

The appropriate average here is the geometric mean and not the arithmetic mean. The arithmetic mean of 5, 8, 77 is 30 but this is not the correct answer. Correct answer shall be obtained if we calculate geometic mean.

% Rise	*X* *Price at the end of the year taking preceding year as 100*	*log X*
5	105	2.0212
8	108	2.0334
77	177	2.2480
		Σ log X = 6.3026

$$\text{G.M.} = \text{A.L.}\left(\frac{\sum \log X}{N}\right) = \text{AL}\left(\frac{6.3026}{3}\right) = \text{A.L. } (2.1009) = 126.2$$

The average increase from 1996 to 1998 = 126.2 – 100 = 26.2% or approx. 26%. Verification. When the average rise is 30%.

	Year	*Rate of change*	*Total change*	*Price at the end of each year*
I	year	30% on 100	30	130.0
II	year	30% on 130	39	169.0
III	year	30% on 169	50.7	219.7

when the average rise is 26%

I	year	26% on 100.00	26.00	126.00
II	year	26% on 126.00	32.76	158.76
III	year	26% on 158.76	41.28	200.04

When the rise is of 5, 8 and 77% the changed price at the end of each year:

I	year	5% on 100.00	5.00	105.00
II	year	8% on 126.00	8.40	113.40
III	year	77% on 158.76	87.318	200.00

The above calculations make it clear that in the second and third cases the price at the end of the third year is almost the same, the slight difference being due to approximation of 26.2 to 26. Hence, the average increase is 26%.

Example 37:

The geometric mean of 10 observations was calculated as 28.6. It was later discovered that one of the observations was recorded as 23.4 instead of 32.4. Apply appropriate correction and calculate the correct geometric mean.

Solution:

Geometric mean of n observations is given by:

$$\text{G.M.} = (X_1, X_2, X_3, \ldots, X_{n)})^{1/n}$$

$$\text{or G.M.}^n = X_1, X_2, X_3, \ldots X_n$$

Thus the product of the number is given by G.M^n or $(28.6)^{10}$ since in the given case n = 10 and G.M. = 28.6.

If the wrong observation 23.4 is replaced by the correct value 32.4, the correct value of the product of 10 numbers is obtained by dividing $(28.6)^{10}$ by wrong observation and multiplying by the correct observation. Hence

$$\text{Corrected product } (X_1, X_2, \ldots, X_n) = \frac{(26.6)^{10} \times 32.4}{23.4}$$

Correct value of geometric mean is given by:

$$\text{G.M.c} = \left[\frac{(28.6)^{10} \times 32.4}{23.4}\right]^{1.10}$$

$$\log \text{G.M.}_c = \frac{1}{10}[10 \log 28.6 + 32.4 - \log 23.4]$$

$$= \frac{1}{10} [10 \times 1.4564 + 1.5105 - 1.3692]$$

$$= \frac{1}{10}[14.564 + 1.5105 - 1.3692] = 1.470\text{-}53$$

$$G.M._c = A.L.\ 1.47053 = 29.54.$$

WEIGHTED GEOMETRIC MEAN

Like weighted arithmetic mean, we can also calculate weighted geometric mean with the help of the following formula:

$$G.M.\ 10 = A.L. \left[\frac{(\log X_1 \times W_1) + (\log X_2 \times W_2) + ... + (\log X_n \times W_n)}{W_1 + W_2 + W_3 + ... + W_n}\right]$$

$$= A.L. \left[\frac{\sum(\log X \times W)}{\sum W}\right]$$

Symbolically, $G.M._w = \sqrt{X_1^{w1} \times X_2^{w2} \times X_3^{w3} \times ... \times X_2^{wn}}$

If W_1, W_2, W_3,..., W_n are weights assigned to different values of X_1, X_2, X_3,..., X_n

$$G.M._w = \sqrt{X_1^{w1}, X_2^{w2}; X_3^{w3} ... X_n^{wn}}$$

Taking logaritms of both sides

$$\log G.M._w = \frac{\log\left(X_1^{w1}, X_2^{w2}, X_3^{w3} ... X_n^{wn}\right)}{\sum W}$$

where $N = \sum W = W_1, + W_2, + W_3 + ..., + X_n$

$$= \frac{\log\left(X_1^{w1} + \log X_2^{w2} + \log X_3^{w3} + ... \log X_n^{wn}\right)}{\sum W} = \frac{\sum W \log X}{\sum W}$$

$$G.M._w = \text{Antilog}\ \frac{\sum W \log X}{\sum W}$$

Since it is difficult to find nth root, the Geometric Mean can be calculated with the help of Logarithms.

Example 38:

The annual rates of growth of an economy over the last five years were 1.5, 2.7, 3.0, 4.5 and 6.2 percent respectively. What is the compound rate of growth perannum of the economy for the period?

Solution:

Apply the geometric mean.

Annual rate of growth	*Growth relatives at the end of the year X*	*log*
1.5	101.5	20.0064
2.7	102.7	2.0116
3.0	103.0	2.0128
4.5	104.5	2.0191
6.2	106.2	2.0261
		Σ log X = 10.076

$$\text{G.M.} = \text{AL.}\left(\frac{\sum \log X}{N}\right) = \text{AL}\left(\frac{10.076}{5}\right) = \text{A.L. } (2.0152) = 103.5$$

The compound rate of growth per annum = 103 – 100 = 3.5

Note: The same result shall be obtained by applying the compound interest formula which is discussed below.

COMPOUND INTEREST FORMULA

Geometric mean is most frequently used in the determination of average per cent of change. For example, if a city had a population of 2,00,000 in a given year and 2.40.000 ten years later, we may be interested in finding out the annual per cent of change. The increase is (2,40,000 – 2,00,000), *i.e.*, 40,000 over a period of 10 years and so one may say that the annual per cent increase is 2. However, if we compute 2 per cent increase each year over the preceding year the population figure turns out to be 2,43, 800. This means that the correct figure is little less than 2 per cent because we are actually compounding. The average annual per cent increase may be computed by applying the formula:

$$P_n = P_0(1 + r)_n$$

where P_0 = The value at the beginning of the period;

P_n = The value at the end of the period n;

r = Rate of change;

n = length of time period.

It follows from the above formula that $r = n\sqrt{\frac{P_n}{P_0}} - 1;$

For the above data 2,40,000 = 2,00,000 $(1 + r)^{10}$

Taking logarithms $5.3802 = 5.301 + 10 \log (1 + r)$

$$\log (1 + r) = 0.00792$$

$$(1 + r) = \text{Antilog } 0,0079$$

$$1 + r = 0.0184 \text{ or } r = 0.0184 = 1.84 \text{ per cent.}$$

The expression $P_n = P_0 (1 + r)^n$ is called compound interest formula because it is extremely useful in problems involving compound interest. In the above case we have used it to determine average annual per cent of growth. However, if we know any three values of the four used in the formula we can find out the fourth one. Thus, we may determine:

(a) Average annual per cent of change r.

(b) Population the given number of years later Pn assuming the constant relative change.

(c) Number of years, n, after which the given population will be attained, again assuming a constant relative change.

(d) Population the given number of years earlier P_0, if the per cent of change was constant.

It may by pointed out that the assumption of a constant relative change for population is not valid over extended period for any country except possibly "new" countries.

Example 39:

The weighted geometric mean of the four numbers 8, 25, 17 and 30 is 15.3. If the weights of the first three numbers are 5, 3 and 4 respectively, find the weight of the fourth number.

Solution:

Let the weight of the fourth number be W_1.

Calculation of Geometric Mean

X	*W*	*Log X*	*W. Log X*
8	5	0.9031	4.5155
25	3	1.3979	4.1937
17	4	1.2304	4.9216
30	W1	1.4771	$1.4771 + W_1$
	$\Sigma W = 12 + W_1$		$\Sigma W.$ Lot $X = 13.6308$
			$1.4771 + W_1$

$$\text{Log G.M.}_w \left[\frac{\sum W \log X}{\sum W}\right] \log 15.3 = \frac{13.63.08 + 1.4771\, W_1}{12 + W_1}$$

$$1.1847\,(12 + W_1) = 13.6308 + 1.4771\, W_1$$

$$14.2164 + 1.1847\, W_1 = 13.6308 + 1.4771\, W_1$$

$$1.1847\, W_1 - 1.471\, W_1 = 13.6308 - 14.2164 = -0.5856$$

$$W_1 = \frac{0.5856}{0.2924} = 2.003 \text{ or } 2 \text{ app.}$$

Thus the weight of the fourth number is 2.

Merits and Limitations of Geometric Mean

Merits

It is based on each and every item of the series.

(a) It gives less weight to large items and more to small ones than does the arithmetic average. It is because of this reason that geometric mean is never larger than the arithmetic mean. On occasions it may turn out to be same as the arithmetic mean, but usually it is smaller.

(b) It is useful in average ration and percentages and in determining rates of increase and decrease.

(c) It is rigidly defined.

(d) It is capable of alebratic manipulation. For example, if the geometric average of two or more series and their number of items is known, a combined G.M. can be easily calculated by applying the formula.

$$\text{G.M.}_{12} = \text{antilog}\left[\frac{N_1 \log \text{G.M.}_1 + N_2 \log \text{G.M.}_2}{N_1 + N_2}\right]$$

For example if there are 2 sets of two figures each and their geometric means are 8 and 12 respectively, we can calculate the combined geometric mean as follows:

$$\text{G.M.}_{12} = \text{A.L.}\left[\frac{(2 \times \log 8)\,(2 \times \log 12)}{2 + 2}\right]$$

$$= \text{A.L.}\left[\frac{(2 \times 9031)\,(2 \times 1.0792)}{4}\right]$$

$$= \text{A.L.}\left[\frac{1.8062 + 2.1584}{4}\right] = \text{A.L. } 0.99115 = 9.8$$

Limitations

It is difficult to understand.

(a) It cannot be computed when there are both negative and positive values in a series or one or more of the values are zero.

(b) It is difficult to compute and to interpret and so has restricted application.

HARMONIC MEAN

The harmonic mean is based on the reciprocals of numbers averaged. It is defined as the reciprocal of the arithmetic mean of the reciprocal of the individual observations. Thus, by definition

$$\text{H.M.} = \frac{N}{\left(\frac{1}{X_1} + \frac{1}{X_2} + \frac{1}{X_3} + \ldots + \frac{1}{X_n}\right)}$$

When the number of items is large, the computation of harmonic mean in the above manner becomes tedious. To simplify calculations we obtain reciprocal of the various items from the table and apply the following formulae:

In individual observations. $\text{H.M.} = \dfrac{N}{\sum(1/X)}$

In discrete series, $\text{H.M.} = \dfrac{N}{\sum\left(f \times \frac{1}{X}\right)}$

In continuous series $\text{H.M.} = \dfrac{N}{\sum\left(f \times \frac{1}{X}\right)} = \dfrac{N}{\sum(f/m)}$

Calculation of Harmonic Mean–Individual Observations

In individual series harmonic mean is computed by applying the following formula:

$$\text{H.M.} = \frac{N}{\left(\frac{1}{X_1} + \frac{1}{X_2} + \frac{1}{X_3} + \ldots + \frac{1}{X_n}\right)}$$

X_1, X_2, X_3, etc, refer to the various items of the variable.

Example 40:

An automobile driver travels from plain to hill station 1000 km, distance at an average speed of 30 km. per hour. He then makes the return trip at average speed of 20 km. per hour. What is his average speed over the entire distance (200 km)?

Solution:

If the problem is given to a layman he is most likely to compute the arithmetic mean of two speeds *i.e.*,

$$\overline{X} = \frac{30 \text{ km} + 20 \text{ km.}}{2} = \text{km. ph.}$$

But this is not the correct average. Harmonic mean would be more suitable in this situation. Harmonic mean of 30 and 20 is

$$\text{H.M.} = \frac{2}{\frac{1}{20} + \frac{1}{30}} = \frac{2}{\frac{10}{120}} = \frac{2 \times 120}{20} = 24 \text{ km.p.h.}$$

It can be pròved that harmonic mean is the appropriate average in this case by tabulating the time and distance for each trip separately as follows:

	Distance (km)	***Average speed km. p.h.***	***Time taken***
Ging	100	30	3 hours 20 minutes
Returning	100	20	5 hours
Total	200		8 hours 20 minutes

Thus the total time required for covering a distance of 200 km. is 8 hours 20 minutes which gives an average speed of 24 km. p.h. and not 25 km. p.h.

The above problem can be changed in such a manner that arithmetic mean is the appropriate average. Suppose the driver makes the same trip but it is given that he travels at 30 km. per hour for half of the time and at 20 km. per hoür for other half of the time Now the correct answer about the average speed would be given by the arithmetic mear *i.e.* average speed = (30 + 20)/2 = 25 km. per cent To verify the result we again prepare a table of time and distance at each speed:

Speed km. p.h	***Distance***	***Time required***
30	120 km.	120/30 = 4 hours
20	80 Km.	80/20 = 4 hours
Total	200 Km.	8 hours

Thus he has covered 200 km. in 8 hours. Hence the average speed is 25 km. per hour.

The above example clearly shows that when distances are the same for the two speeds harmonic mean gives the correct answer but when times are same, the arithmetic mean of the rates of speed gives the correct answer.

Example 41:

From the following data compute the value of harmonic mean:

Class interval	*10-20*	*20-30*	*30-40*	*40-50*	*50-60*
Frequency	*4*	*6*	*10*	*7*	*3*

Solution:

Calculation of Harmonic Mean

Class interval	***Mid-points m***	***Frequency f***	***f/m***
10-20	15	4	0.267
20-30	25	6	0.240
30-40	35	10	0.286
40-50	45	7	0.156
50-60	55	3	0.055
		N = 30	Σ(f/m) = 1.004

$$\text{H.M.} = \frac{N}{\Sigma(1/X)} = \frac{30}{1.004} = 29.88.$$

Uses of Harmonic Mean

The harmonic mean is restricted in its field of usefulness. It is useful for computing the average rate of increase in profits of a concern or average speed at which a journey has been performed or the average price at which an article has been sold. The rate usually indicates the relation between two different types of measuring units that can be expressed reciprocally. For example if a man walked 20 km. in 5 hours the rate of his walking speed be expressed as

$$\frac{20 \text{ km.}}{5 \text{ hours}} = 4 \text{ km. per hour}$$

where the unit of the first term is a km. and the unit of the second term is an hour. Or reciprocally,

$$\frac{20 \text{ km.}}{5 \text{ hours}} = \frac{1}{4} \text{ hours per km.}$$

where the unit of the first term is an hour and the unit of the second is term is km.

Example 42:

From the following data compute the value of harmonic mean:

Marks	*10*	*20*	*25*	*40*	*50*
No. of students	*20*	*30*	*50*	*15*	*5*

Solution:

Calculation of Harmonic Mean

Marks X	***f***	***(f/X)***
10	20	2.000
20	30	1.500
25	50	2.000
40	15	0.375
50	5	0.100
	N = 120	Σ(f/X) = 5.975

$$\text{H.M.} = \frac{N}{\sum(1/X)} = \frac{120}{5.975} = 20.08.$$

Calculation of Harmonic Mean–Continuous Series

For calculating harmonic mean in continuous series the procedure is the same as applied to discrete series. The only difference is that here we take the reciprocal of the mid-points.

Example 43:

An aeroplane covers the four sides of a square at speeds of 1,000 2,000 3,000 and 4,000 km. per hour respectively. What is the average speed of the plane in its flight around the square?

Solution:

If we compute the arithmetic mean we get the following answer;

$$\overline{X} = \frac{1{,}000 + 2{,}000 + 3{,}000 + 4{,}000}{4} = \frac{10{,}000}{4} = 2{,}500 \text{ km. per hour.}$$

However that is not the correct answer. In such a problem harmonic mean is an appropriate average

$$\text{H.M.} = \frac{4}{\frac{1}{1{,}000} + \frac{1}{2{,}000} \; \frac{1}{3{,}000} + \frac{1}{4{,}000}}$$

$$= \frac{4}{\frac{12+6+4+3}{12{,}000}} = \frac{4}{\frac{25}{12{,}000}} = \frac{4 \times 12{,}000}{25} = 1{,}920 \text{ km. per hour.}$$

Verification. Suppose one side of square is 1000 km. Each side *i.e.* 1,000 km. it covers at average speeds of 1,000 2,000 3,000 4,000 kms respectively. From this we can calculate the time taken in covering the entire distance.

Distance	***Speed (km. p.h)***	***Time taken***
1,000	1,000	60 minutes
1,000	2,000	30 minutes
1,000	3,000	20 minutes
1,000	4,000	15 minutes
Total		125 minutes

In 125 minutes it covers 4,000 km.

In 60 minutes is would cover $\frac{4{,}000}{125} = 60 = 1{,}920$ km.

Thus the average speed over the entire distance is 1,920 and not 2,500 km. per hour.

WEIGHTED HARMONIC MEAN

At times it may be necessary to calculate the weighted harmonic mean. For example if we are given not only the speed of the aeroplane but the distances travelled also simple harmonic mean cannot be used. Weighted harmonic mean is calculated with the help of the following formula:

$$\text{H.M.}_{\omega} = \frac{\sum \omega}{\left(\frac{1}{a} \times \omega_1\right) + \left(\frac{1}{b} \times \omega_2\right) + \left(\frac{1}{c} \times \omega_3\right)} \text{ or } \frac{\sum \omega}{\sum(\omega / x)}.$$

Example 44:

Find the weighted geometric mean from the following data:

Group	*Index Number*	*Weights*
Food	*260*	*46*
Fuel & Lighting	*180*	*10*
Clothing	*220*	*8*
House Rent	*230*	*20*
Education	*120*	*12*
Misc.	*200*	*4*

Solution:

Calculation of Weighted Geometric Mean

Group	*Index No. X*	*Weights W*	*Log X*	*W log X*
Food	260	46	2.4150	111.0900
Fuel & Lighting	180	10	2.2553	22.5530
Clothing	220	8	2.3424	18.7392
House Rent	230	20	2.3617	47.2340
Education	120	12	2.0792	24.9504
Misc.	200	4	2.3010	9.2040
		$\Sigma W = 100$		$\Sigma W \text{ Log } X = 233.7706$

$$G.W._{w} = A.L.\left[\frac{\Sigma W \log X}{\Sigma W}\right] = A.L.\left[\frac{233.7706}{100}\right] = A.L\ 2.3377 = 217.6$$

Example 45:

(a) Find the harmonic mean from the following:

2574 475 75 5 0.8 0.08 0.005 0.0009.

Solution:

Calculation of Harmonic Mean

X	*(1/X)*	*X*	*(1/X)*
2574	0.0004	0.8	1.2500
475	0.0021	0.08	12.5000
75	0.0133	0.005	200.0000
5	0.2000	0.0009	1111.1111
			$\Sigma(1/X) = 1325.0769$

$$H.M. = \frac{N}{\Sigma(1/X)} = \frac{8}{1325.0769} = 0.006.$$

Example 46:

(b) Calculate the harmonic mean from the following data:

3834 382 63 8 0.4 0.03 0.009 0.005

Solution:

Calculation of Harmonic Mean

X	*(f/X)*	*X*	*(f/X)*
3834	0.0003	0.4	2.5000
382	0.0027	0.03	33.3333
63	0.0159	0.009	111 1111
8	0.1250	0.0005	2000.0000
	N = 120		Σ(1/X) = 5.975

$$H.M. = \frac{N}{\Sigma(1/X)} = \frac{8}{2147.0883} = 0.003726.$$

Calculation of Harmonic Mean-Discrete Series

In discrete series, harmonic mean is computed by applying the following formula:

$$H.M. = \frac{N}{\Sigma\left(f \times \frac{1}{X}\right)} = \frac{N}{\Sigma(f/m)}$$

(i) Take the reciprocal of the various items of the variable X.

(ii) Multiply the reciprocal by frequents and obtain the total $\Sigma\left(f \times \frac{1}{X}\right)$.

(iii) substitute the values of N and $\Sigma\left(f \times \frac{1}{X}\right)$. in the above formula.

Note: Instead of first finding out the reciprocals and then multiplying them by frequencies it will be far more easider to divide each frequency by the respective value of the variable.

Example 47:

A cycilist covers his first five km. at an average speed of 10 km. p.h. another km. at 8km. p.h. and the last two km. at 5 km. p.h. Find average spped of the entire journey and verify your answer.

Solution:

Weighted harmonic mean would be appropriate here

$$H.M._w = \frac{\Sigma w}{\left(\frac{1}{a} \times w_1\right) + \left(\frac{1}{b} \times w_2\right) + \left(\frac{1}{c} \times w_3\right)}$$

The various speeds are 10 km. ph. 8 km. p.h. and 5 km. henceab and c respectively are 108 and 5. The distances covered are respectively 53 and 2 km. Hence w_1, w_2 and w_3 3 are 53 and 2

$$H.M._w = \frac{10}{\left(\frac{1}{10} \times 5\right) + \left(\frac{1}{8} \times 3\right) + \left(\frac{1}{2} \times 2\right)}$$

$$= \frac{10}{\frac{1}{2} + \frac{3}{8} + \frac{2}{5}} = \frac{10}{\frac{51}{40}} \; \frac{10 \times 40}{51} = 7.84. \text{ km. p.h.}$$

Thus the average speed for the entire journey is 7.84 km. p.h.

Verification:

Speed (km. p.h.)	***Distance (km.)***	***Time taken (in minutes)***
10	5	30.0
8	3	22.5
5	2	24.0
Total	10	76.5

In 76.5 minutes he covers 10 km.

In 1 minute the would cover $\frac{10}{76.5} \times 60 = 7.84.$

In 60 minutes he would cover $\frac{10}{76.5} \times 60 = 7.84$ km.

Example 48:

The following figures show the number of passengers carried on each of 50 journeys by an aircraft with a seating capacity of 100. Calculate:

(i) the average capacity used and (ii) it 65 passengers is the smallest profitable load the proportion of lights which were unprofitable.

10	*18*	*61*	*63*	*72*	*71*	*82*	*25*	*45*	*66*	*68*	*95*	*92*
31	*41*	*56*	*33*	*78*	*65*	*72*	*74*	*89*	*67*	*68*	*32*	*46*

49	35	37	43	55	57	69	72	84	69	92	75	83
68	75	42	45	39	11	37	38	62	89	72		

Solution:

(i) Let us obtain the total of all the items given and find out the arithmetic mean

$$\overline{X} = \frac{\Sigma X}{N} = \frac{2918}{50} = 58.36$$

Thus the average capacity used is 58.36 passengers.

(ii) The number of flights that carried passengers below 65 is 23. Hence the proportion of flights which were unprolitable would be 23/50 × 100 = 46 per cent.

Merits and Limitations of Harmonic Mean

Merits

Its value is based on every item of the series.

(a) In problems relating to time and rates it gives better results than other averages.

(b) It tends itself to algebraic manipulation.

Limitations

It is not easily understood.

(a) It gives largest weight to smallest items. This is generally not a destrable feature and as such this average is not very useful for the analysis of economic data.

(b) It is difficult to compute.

(c) Its value cannot be computed when there are both positive and negative items in a series or when one or more items are zero.

Because of these limitations the harmonic mean has little practical application and is not a good representation of a statistical seriousness the phenomenon is such where small items need to be given a very high weightage.

RELATIONSHIP AMONG THE AVERAGES

In any distribution when the original items differ in size the value of A.M. G.M. and H.M. would also differ and will be in the following orr:

$$\text{A.M.} \geq \text{G.M.} \geq \text{H.M.}$$

i.e. arithmetic mean is greater than geometric mean and geometric mean is greater than harmonic mean. The equality signs hold only if all the numbers X_1, X_2,... X_n are identical.

Proof:

Prove that if a and b are two positive numbers their A.M. ≥ G.M. ≥ H.M.

Let a and b be two positive quantities such that a 1 b.

The A.M and H.M. of these two quantities are

$$\overline{X} = \frac{a+b}{2}; \text{ G.M.} = \sqrt{a \times b}; \text{ H.M.} = \frac{2}{\frac{1}{a}+\frac{1}{b}} = \frac{2ab}{a+b}$$

We have to prove that A.M. > G.M > H.M. Let us first prove that A.M. > G.M. or $\frac{a+b}{2} > \sqrt{a+b}$.

$$\frac{a+b}{2} > \sqrt{a+b}; a+b > \sqrt{ab}$$

$$a+b-2\sqrt{ab} > 0 \qquad [\text{Since } a+b-2\sqrt{ab} = (\sqrt{a}-\sqrt{b})^2]$$

$$(\sqrt{a}-\sqrt{b})^2 > 0.$$

But the square of any real quantity is positive. Hence $(\sqrt{a}-\sqrt{b})^2$ will be positive. Hence $\frac{a+b}{2} > \sqrt{ab}$

Let us now prove that G.M. > H.M.

$$\Rightarrow \qquad \sqrt{ab} > \frac{2ab}{a+b} \; 1 > \frac{\sqrt{ab}}{a+b} \text{ or } a+b > 2\sqrt{ab}.$$

This has slready been proved above. Hence G.M. > H.M.

Since we have shown that A.M. > G.M. and G.M. > H.M. it is automatically proved that A.M. > G.M. > H.M.

If a and b are equal in that case A.M. = G.M. = H.M. Thus,

A.M. > G.M. > H.M.

MISCELLANEOUS EXAMPLES

Example 1:

In 500 small-scale industrial units the return on investment ranged from 0 to 30 per cent, no unit sustaining any loss, 5 per cent of the units had returns ranging from 0 per cent up to (and including) 5 per cent and 15 per cent of the units earned returns cent and the upper quartile 20 per cent. The uppermost layer of the returns exceeding 25 per cent was earned by 50 units.

Present this information in the form of a frequency table with intervals of 5 per centas follows:

Exceeding 0 per cent but not exceeding 5 per cent

"	5	"	"	10	"
"	10	"	"	15	"
"	15	"	"	20	"
"	20	"	"	25	"
"	25	"	"	30	"

Use $\frac{N}{4}, \frac{2N}{4}, \frac{3N}{4}$ *as the ranks of the tower middle and upper quartiles respectively. Find the rate of return round which there is maximum concentration of the units.*

Solution:

The information given above can be summarized as follows:

Table Showing the Distribution of Small-Scale Industrial Units According to the Table of Returns of Investment

Rate of Return on Investment Number of			***Firms % of***		
		Total			**Firms**
Exceeding 0 but not exceeding	5		5	(a)	25
5	10		15	(b)	75
10	15	(b)	30	(d)	150
15	20	(c)	25	(f)	125
20	25	(b)	15	(h)	75
25	30		10		50
			100		500

On the basis of information provided the computations are shown below:

(a) 5% of 500 = 25,

(b) 15% of 500 = 75

(c) The rate of return of 15% being the median 3N/2 would represent 50% of the firms 20% of the data is comprised in previous class. Tusclasses would represent 50% of the firms 20% of the data is comprised in previous calss. Thus classes would represent 30% of the firms.

(d) In the consequence 30% of 500 = 150.

(e) Firms having 20% return constitute the quartile $\frac{3N}{4}$ This shall cover 75% of the data comprised in the preceding calss being representation of 50% firms this class will cover 25% of data.

(f) In consequence 25% of 500 = 125.

(g) The residual balance of given data equals 15%.

(h) 15% being the residual balance it represents 75% firms.

The rate of return around which there is maximum concentration is the modal calss. The modal lies in the class 10-15.

$$\text{Mode} = L + \frac{\Delta_1}{\Delta_1 + \Delta_2} \times i$$

$$L = 10 \ \Delta_1 = 150 - 75 = 75$$

$$\Delta_2 = 150 - 125 = 25 \ i = 5$$

$$\therefore \quad M_0 = 10 + \frac{75}{75 + 25} \times 5 + 10 + 3.75 = 13.75$$

Hence the rate of return around which there is maximum concentration of units is 13.75%.

Example 2:

The number of fully formed apples on 100 plants were counted with following results:

2	*Plants*	*had*	*0*	*apples*
5	"	"	*1*	"
7	"	"	*2*	"
11	"	"	*3*	"
18	"	"	*4*	"
24	"	"	*5*	"
12	"	"	*6*	"
8	"	"	*7*	"
6	"	"	*8*	"
4	"	"	*9*	"
3	"	"	*10*	"

(i) How many apples were there in all?

(ii) What was the average of number of apples per plant?

(iii) What was the modal no. of apples?

Solution:

Calculation of Total Number average Number and Modal Number of Apples

X	*f*	*fX*
0	2	0
1	5	5
2	7	14
3	11	33
4	18	72
5	24	120
6	12	72
7	8	56
8	6	48
9	4	36
10	3	30
	N = 100	ΣfX = 486

(i) Total number of apples = 486

(ii) Averages no. of apples $= \frac{\Sigma fX}{N} = \frac{486}{100} = 4.86$

(iii) Since the highest frequency is 24, the modal number of apples was. 5.

Example 3:

The following are the weekly wages in rupees of 30 workers of a firm:

140	*139*	*126*	*114*	*100*	*88*	*62*	*77*	*99*	*103*	*108*
129	*144*	*148*	*134*	*63*	*69*	*148*	*132*	*118*	*142*	*116*
123	*104*	*95*	*80*	*85*	*106*	*123*	*133*			

The firm gave bonus of Rs. 10 15 20 25 30 and 35 for individuals in the respective salary: exceeding 60 but not exceeding 75 exceedings but not exceeding 90 and so on up to exceeding 135 and not exceeding 150.

Find the average bonus paid.

Solution:

Let us first prepare a frequency distribution of the given data and then calculate the average bonus paid.

Weekly wages (Rs.)	*Tally Bars*	*Frequency f*	*Bonus Paid X*	*fx*
61-75	|||	3	10	30
76-90	||||	4	15	60
91-105	||||	5	20	100
106-120	||||	5	25	125
121-135	|||| ||	7	30	210
136-150	||||		6	35	210
		n = 30		ΣfX = 735

$$\text{Average bonus paid} = \frac{\Sigma X}{N} = \frac{735}{30} \text{ Rs. } 24.5.$$

Example 4:

Calculate arithmetic mean median and mode from the following frequency distribution:

Variable	*Frequency*	*Variable*	*Frequency*
10-13	*8*	*25-28*	*54*
13-16	*15*	*28-31*	*36*
16-19	*27*	*31-34*	*18*
19-22	*51*	*34-37*	*9*
22-25	*75*	*37-40*	*7*

Solution:

Calculation of Mean Median & Mode

Variable	*m.p. m*	*f*	*(m–23.5)/3 d*	*fd*	*c.f*
10-13	11.5	8	–4	–32	8
13-16	14.5	15	–3	–45	23
16-19	17.5	27	–2	–54	50
19-22	20.5	51	–1	–51	101
22-25	23.5	75	0	0	176
25-28	26.5	54	+1	+54	230
28-31	29.5	36	+2	+72	266
31-34	32.5	18	+3	+54	284
34-37	35.5	9	+4	+36	293
37-40	38.5	7	+5	+35	300
		N = 300		Σfd = 69	

$$\overline{X} = A + \frac{\Sigma X}{N} i = 23.5 + \frac{69}{300} \times 3 = 24.19$$

Med. = Size of N/2th item = 300/2 = 15th item

Median lies in the class 22 – 25.

$$\text{Med.} = L + \frac{N/2 - c.f.}{f} \times i = 22 + \frac{150 - 101}{75} \times 3 = = 22 + 1.96 = 23.96$$

Mode. By inspection mode lies in the class 22 – 25.

$$M_0 = L + \frac{\Delta_1}{\Delta_1 + \Delta_2} \times i$$

$$L = 22 = \Delta_1 = (75 - 51) = 24 \ \Delta_2 = (75 - 54) = 21, \ i = 3$$

$$M_0 = 22 + \frac{24}{24 + 21} \times 3 = 22 + 1.6 = 23.6.$$

Example 5:

The mean annual salaries paid to 1,000 employees of a company was Rs. 5,000. The mean annual salaries paid to male and female employees were Rs. 5,200 and Rs. 4,200 respectively. Determine the percentage of males and females employed by the company.

Solution:

Let N_1 represent percentage of males and N_2 percentage of females so that $N_1 + N_2 = 100$

We are given $\overline{X}_{12} - 5,000$ $\overline{X}_1 - 5,200$, $\overline{X}_2 - 4,200$.

Substituting the values in the formula:

$$\overline{X}_{12} = \frac{N_1\overline{X}_1 + N_2\overline{X}_2}{N_1 + N_2}; \ 5,000 = \frac{N_1(5,200) + N_2(4,200)}{100}$$

$$5,00,000 = 5,200 \ N_1 + [(100 - N_1)(4,200)]$$

$$= \left[\begin{array}{c} \text{Since } N_1 + N_2 = 100 \\ N_2 = 100 - N_1 \end{array} \right]$$

$$\Rightarrow \quad 5,00,000 = 5,200 \ N_1 + 4,20,000 - 4,200 \ N_1$$

$$1,000 \ N1 = 80,000$$

$$N_1 = 80$$

and $N_2 = 100 - N_1 \ (100 - 80) = 20$

Thus, the percentage of males and females employed is 80 and 20 respectively.

Which Average to Use

We have explained above the methods of computing the various types of averages and also their distinctive features. At this point the reader has a right to ask "Which of these averages should I use"? or "When ought I to use one or the other of the averages described"? "or "Which of these is the best average to be used"

It must be clearly understood that no one average can be regarded as best for all circumstances. The following considerations influence the selection of an appropriate average:

(a) The type of data available. Are they badly skewed (avoid the mean), gappy around the middle (avoid the median) or unequal in class interval (avoid the mode)?

(b) would the average be sued for further computations?

(c) The typical value required in the particular problem. Within the framework of descriptive statistics the main requirement is to know what each average means and then select one that fulfils the purpose in hand. Is a composite average of all absolute or relative values needed (arithmetic mean or geometric mean) or is middle value wanted (median) or the most common value (mode)?

(d) The purpose which the average is designed to serve.

On occasions it may even be advisable to work out more than one average and present them although to be sure this procedure creates an added burden for the reader as well as for the statistician. But he added burden is preferable to the use of single average that may be an incomplete description. To use it also is like looking through a key hole the part of the room you can see cannot give a full idea of the whole room.

Median

The median is generally the best average in open-end grouped distributions especially where if plotted as a frequency curve one gets a J or reverse J curve; for example in case of price distribution or income distribution. In such cases very high or very low values would cause the mean to be higher to lower than the most "common" values. In such instances the median or middle value of the series may be more representative figure to use in describing the mass of data.

Mode

Generally speaking the need of mode lies in the fact that it can be used to describe quantitative data. The mode can be used in problems involving the expression of preferences where quantitative measurements are not possible.

Thus the preferred type of package design among a number of alternative design would be the modal design. If we want to compare consumer preferences for different kinds of products or different kinds of advertising we can compare the modal preferences expressed by different groups of people but we cannot calculate the median or mean Mode is a particularly useful average for discrete series e.g., number of people wearing a given size of shoe or number of children per household etc. The mode is best suited where there is an outstandingly large frequency.

Geometric Mean

Geometric mean is useful in averaging ratio and percentages and in computing average rates of increase or decrease. It is particularly important in Economics and Business Statistics in index number construction.

Harmonic mean is useful in problems in which of a variable are compared with a constant quantity of another variable *i.e.*, rates time distance covered within certain time and quantities purchased or sold per unit etc.

HARMONIC MEAN

In the following cases arithmetic mean should not be used:

(a) when there are very large and very small items arithmetic mean would be seriously misleading on account of undue influence from extreme items.

(b) The arithmetic mean should not be used to average ration and rates of change. In such cases the geometric mean is more suitable.

(c) When the distribution is unevenly spread concentration being smaller or large at irregular points.

(d) In distributions with open-end intervals.

(e) In highly skewed distributions.

Leaving aside the above specific cases where either median mode geometric mean or harmonic mean is more appropriate in other cases we should apply as a rule of thumb the arithmetic mean–the most popular and widely used average in practice.

It may also be pointed out that a complete description of distribution occasionally calls for two or more of these averages. It is true that presenting two or more averages creates an added burden for the investigator as well as for the consumer of statistics. However, the work this extra burden entails is fully justified if it presents a more complete description of the data that is possible from a single measure.

General Limitations of Average

(a) At times the average may give a very absurd result. For example if we are calculating size of a family we may get a value 4.8. But this is impossible as persons cannot be in fractions. However we should remember that it is an average value representing the entire group.

(b) An average may given us value that does not exist in the data. For example the arithmetic mean of 100, 300, 250, 50, 100 is = 160 a value that does not exist in the data.

(c) Measure of central value fall to give an idea about the formation of the series. Two or more series may have the same central value but may differ widely in composition. For example observe the following two series:

Series A	*Series B*
150	300
170	500
190	20
210	20
180	2
Total 900 $\overline{X}$ = 180	900 = 180

(d) Since an average is a single value representing a group of values it must be properly interpreted; otherwise there is every possibility of Jumping to wrong conclusions. This can best illustrated with the help of a story. A person had to cross the river from one bank to another. He was not aware of the depth of the river so he enquired of another man who told him that the average depth of water is 5' 4". The man was 5' 6" and the he though that he can very easily cross the river because at all time he would be above the level of water. So he started. In the beginning the level of water. So he reached the middle the water was 15 ft. deep and he lost his life. The man was drowned because he had a misconception that average depth means uniform depth throughout. But it is not so. An average represents a group of values and lies somewhere in between the tow extremes *i.e.* the largest and the smallest items of the series.

Example 6:

The price of a commodity doubles in a period of 4 years. What is the average percentage increase per year?

Solution:

Geometric mean would be more appropriate here. Applying the following formula:

$$P_n = P_0 \left(1 + \frac{r}{100}\right)^n$$

Here $\quad P_n = 200 \; P_0 = 100$ and $n = 4$

$$200 = 100 \left(1 + \frac{r}{100}\right)^n$$

$$\left(1 + \frac{r}{100}\right) = 2^{1/4}$$

Let $\quad X = 1^{1/4}$

$$\text{Log } X = \frac{1}{4} \text{ Log } 2 \; \frac{1}{4} \; (0.301) = 0.07525$$

$$X = \text{AL } (0.07525) = 1.19$$

$$\left(1 + \frac{r}{100}\right) = 1.19 = 1.19 - 1 = 0.19 \text{ or } 19 \text{ per cent.}$$

$$\frac{r}{100} = 1.19 - 1 = 0.19 \text{ or } 19 \text{ per cent.}$$

Example 7

The arithmetic mean the mode and the medium of a group of 75 observations were calculated to be 27, 34 and 29 respectively. It was later discovered that one observation was wrongly read as 43 instead of the correct value 53. Examine to what extent the calculated values of the three averages will be affected by error.

Solution:

Correct mean $\quad \overline{X} = \frac{\Sigma X}{N}$ or $\Sigma X = N\overline{X}$

$$N = 75 \quad \overline{X} = 27$$

$\therefore \quad \Sigma X = 75 \times 27 = 2{,}025$

Correct $\quad \Sigma X = 2{,}025 - 43 + 53 = 2{,}035$

Correct $\quad \overline{X} = \frac{2{,}035}{75} = 27.13$

The values of median and mode will not be affected by this error because these are positional averages and the median value is 29 which is far away from the values 43 and 53. Similarly the value of mode would not be affected by the error since the modal value 34 is far away from the value 43 and 53.

Example 8:

Calculate Median and Mode of the data given below. Using them find arithmetic mean:

Marks:	*10*	*20*	*30*	*40*	*50*	*60*
No. of students:	*8*	*23*	*45*	*65*	*75*	*80*

Solution:

Calculation of Median, Mode and Arithmetic Mean

Marks	***f***	***c.f.***
0–10	8	8
10–20	15	23
20–30	22	45
30–40	20	65
40–50	10	75
50–60	5	80
	N = 80	**N = 80**

Median

$$\text{Median} = \text{size of } \frac{N}{2} \text{ item} = \frac{80}{2} = 40\text{th item}$$

Median lies in the class 20–30.

$$\text{Med} = L + \frac{N/2 - c.f.}{f} \times i$$

$$L = 20,\ N/2 = 40,\ c.f. = 23,\ f = 22,\ i = 10$$

$$\text{Med} = 20 + \frac{40 - 23}{22} = \times 10 = 20 + 7.73 = 27.73$$

Mode

By inspection mode lies in the class 20–30.

$$Mo = L + \frac{\Delta_1}{\Delta_1 + \Delta_2} \times i$$

$$L = 20,\ \Delta_1 = (22 - 15) = 7,\ \Delta_2 = (22 - 20) = 2,\ i = 10$$

$$Mo = 20 + \frac{7}{7 + 2} \times 10 = 20 + 7.78 = 27.78.$$

Arithmetic Mean

Since we are asked to calculate arithmetic mean using median and mode we will apply the following formula:

Mode = 3 Median – 2 Mean

Substituting the values of mode and median

27.78 = 3 (27.73) – 2 Mean

$$-2\overline{X} + 83.19 = 27.78$$

$$-2\overline{X} = 27.78 - 83.19 = -55.41$$

$$\overline{X} = 27.705$$

Example 9:

10 percent of the workers in a firm employing a total of 1,000 workers earn less than Rs. 5 per day 200 earn between Rs. 5 and 9.99 30 percent between 10 and 14.99 250 workers between 15 and 19.99 and the rest 20 and above. What is the median wage?

Solution:

First convert the given figure in a frequency distribution as follows:

Wages (Rs.)	*No. of workers f*	*c.f.*	
Below 5	1,000 × 10/100 = 100		100
5–9.99	200		300
10–14.99	1,000 × 30/100 = 300	1,000	600
15–19.99	250		850
20 and above	150		1,000

$$\text{Median} = \text{size of } \frac{N}{2}\text{th item} = \text{Size of } \frac{1{,}000}{2} = 500\text{th item}$$

Median lies in the class 10–14.99. But the real limit of this class is 9.995–14.995.

$$\text{Med.} = L + \frac{N/2 - c.f.}{f} \times i$$

$$L = 9.995,\ N/2 = 500,\ c.f. = 300,\ f = 300,\ i = 5$$

$$\text{Med.} = 9.995 + \frac{500 - 300}{300} \times 5 = 9.995 + 3.333 = \text{Rs. } 13.328.$$

Example 10:

A factory pays workers on piece rate basis and also a bonus to each worker on the basis of individual output in each quarter.

The rate of the bonus payable is as follows:

Output in units	**Bonus in Rs.**	**Output in units**	**Bonus in Rs.**
70–74	40	90–94	70
75–79	45	95–99	80
80–84	50	100–104	100
85–89	60		

The individual output of a batch of 50 workers is given below:

94	83	78	76	88	86	93	80	91	82
89	97	92	84	92	80	85	83	98	103
87	88	88	81	95	86	99	81	87	90
84	97	80	75	93	101	82	82	89	72
85	83	75	72	83	98	77	87	71	80

By suitable classification you are required to find:

(i) Average bonus per worker for the quarter.

(ii) Average output per worker.

Solution:

Frequency Distribution By Output & Bonus

Output	*Tallies*	*f*	*Bonus X*	*m.p. m*	*(m-87)/5 d*	*fd*	*fX*
70–74	III	3	40	72	–3	–9	120
75–79	~~IIII~~	5	45	77	–2	–10	225
80–84	~~IIII~~ ~~IIII~~ ~~IIII~~	15	50	82	–1	–15	750
85–89	IIII IIII II	12	60	87	0	0	720
90–94	~~IIII~~ II	7	70	92	+ 1	+ 7	490
95–99	~~IIII~~ I	6	80	97	+ 2	+ 12	480
100–104	II	2	100	102	+ 3	+ 6	200
						Σfd = – 9	ΣfX = 2985

(i) Average bonus per worker for the quarter :

$$\overline{X} = \frac{\Sigma fX}{N} = \frac{2,985}{50} = 59.7$$

(ii) Total quartely bonus paid = 59.7 × 50 = Rs. 2,985

(iii) Average output per worker

$$\overline{X} = A + \frac{\Sigma fX}{N} \times i 87 - \frac{9}{50} \times 5 = 86.1 \text{ units.}$$

Example 11:

From the following data compute the mean marks of all the students of 50 schools in a city.

Marks obtained	*No. of schools*	*Average No. of students in a school*
More than 35	*7*	*200*
30–35	*10*	*250*
25–30	*15*	*300*
20–25	*9*	*200*
15–20	*5*	*150*
Less than 15	*4*	*100*

Solution:

First rewrite the given data in ascending order and then calculate the mean.

Calculation of Mean Marks

Marks	*No. of schools*	*Average No. of students*	*Total No. of students (2 × 3)*	*m.p.m*	*(m-27.5)/5 d*	*fd*
(1)	*(2)*	*(3)*	*(f)*	*m*	*d*	*fd*
10–15	4	100	400	12.5	–3	–1,200
15–20	5	150	750	17.5	–2	–1,500
20–25	9	200	1,800	22.5	–1	–1,800
25–30	15	300	4,500	27.5	0	0
30–35	10	250	2,500	32.5	+1	+2,500
35–40	7	200	1,400	37.5	+2	+2,800
			N = 11,350			Σfd = +800

$$\overline{X} = A + \frac{\Sigma fX}{N} \times i = 27.5 + \frac{800}{11,350} \times 5 = 27.5 + 0.35 = 27.85$$

Example 12:

In a certain examination the average grade of all students in class A is 68.4 and students in class B is 71.2. If the average of both classes combined is 70, find the ratio of the number of students in class A to the number of students in class B.

Solution:

Let us assume that the number of students in class A was 'X' and in class B 'Y'.

We are given $\overline{X}_{12} = 70,\ \overline{X}_1 = 68.4,\ \overline{X}_2 = 71.2$

Substituting these values in the formula:

$$\overline{X} = \frac{N_1\overline{X}_1 + N_2\overline{X}_2}{N_1 + N_2}\ 70 = \frac{68.4\text{ x} + 71.2\text{ y}}{\text{x} + \text{y}}$$

$$70\ (X + Y) = 68.4X + 71.2\ Y$$

$$70X - 68.4X + 70Y - 71.2Y = 0$$

$$\Rightarrow \quad 1.6X = 1.2Y$$

Suppose $X = 10$

$$1.2Y = 16 \text{ or } Y = \frac{16}{1.2} = \frac{40}{3}$$

Thus X and Y are in the ratio of $10 : \frac{40}{3}$ or 30 : 40.

Hence for every 3 students in class A there are 4 students in class B.

Example 13:

Find the class intervals if the arithmetic mean of the following distribution is 33 and assumed mean 35:

Step deviations	*–3*	*–2*	*–1*	*0*	*+1*	*+2*
Frequency	*5*	*10*	*25*	*30*	*20*	*10*

Solution:

Determination of Class Intervals

Step deviations d	***Frequency f***	***fd***
–3	5	–15
–2	10	–20
–1	25	–25
0	30	0
+1	20	+20
+2	10	+20
	N = 100	**Σfd = –20**

$$\overline{X} = A + \frac{\Sigma fX}{N} \times i$$

$$A = 35,\ \overline{X} = 33,\ N = 100,\ \Sigma fd = -20$$

Substituting the values $33 = 35 - \frac{20}{100} \times i$

$$33 - 35 = -0.2\ i$$

$$0.2\ i = 2\ i = \frac{2}{0.2} = 10.\ \text{Thus the class interval is } 10$$

Assumed mean lies in the mid-value of that class '0' as step deviation. The lower and upper limits of this class are:

$35 - \frac{10}{20} = 30$ and $35 + \frac{10}{2} = 40$, *i.e.*, 30–40. The other classes will be:

0–10 10–20 20–30 30–40 40–50 50–60

Note: Since all the step deviations show equal gap, c = i in the formula for calculating arithmetic mean. As we are required to determine class interval we have substituted i in place of c.

Example 14:

Find the numbers whose arithmetic mean is 12.5 and geometric mean 10.

Solution:

Let the numbers of 'a' and 'b'

$$\text{G.M.} = \sqrt{a \times b} = 10 \quad \therefore \quad ab = (10)^2 = 100$$

$$\overline{X} = \frac{a + b}{2} = 12.5$$

$$\therefore \quad a + b = 2 \times 12.5 = 25$$

We know that $(a + b)^2 - (a - b)^2 = 4\ ab$.

Substituting the given values we have

$$(25)^2 - (a - b)^2 = 4 \times 100$$

$$625 - (a - b)^2 = 400$$

$$\Rightarrow \quad (a - b)^2 = 625 - 400 = 225$$

$$a - b = 225 = 15$$

$$a + b = 25$$

$$a - b = 15$$

$$2a = 40 \text{ or } a = 20$$

Substituting the value of 'a' in equation (i).

$$b = 25 - 20 = 5$$

Hence the two numbers are 20 and 5.

Example 15:

The following is the age distribution of 2,000 persons working in a large textile mill:

Age group	*No. of persons*	*Age group*	*No. of persons*
15 but less than 20	*80*	*45 but less than 50*	*268*
20 but less than 25	*250*	*50 but less than 55*	*150*
25 but less than 30	*300*	*55 but less than 60*	*75*
30 but less than 35	*325*	*60 but less than 65*	*25*
35 but less than 40	*287*	*65 but less than 70*	*20*
40 but less than 45	*220*		

Because of the heavy losses the management decides to bring down the strength to 40% of the present number according to the following scheme:

(i) *To retrench the first 10% from lower group.*

(ii) *To absorb the next 40% in other branches.*

(iii) *To mass 10% from the highest age group retire prematurely.*

What will be the age limits of persons retained in the mill and of those transferred to other branches? Also calculate the average age of those retained.

Solution:

The number of persons to be retrenched from the lower group = $\frac{2{,}000 \times 10}{100}$ = 200 = 200. Eighty of these will be from 15–20 age group and the rest (200 – 80) = 120 from 20–25 age group.

The persons to be absorbed in other branches = $\frac{2{,}000 \times 40}{100}$ = 800. They belong to the following age groups:

Age group		*No. of persons*
20–25	(250 – 120)	130
25–30		300
30–35		325
35–40	(287 – 242)	45
		800

Those who are to retire are $\frac{2,000 \times 10}{100}$ = 200 in all and they belong to highest age group. Their age groups are:

Age group		*No. of persons*
65–70		20
60–65		25
55–60		75
50–55	(150 – 70)	80
		200

Hence the age limits of those who are retained in the mill are:

Age group	*No. of persons*
35–40	242
40–45	220
45–50	268
50–55	70
Total	**800**

Calculation of Average Age of Those Retained

Age group	*m.p. m*		*(m – 47.5)5 d*	*fd*
35–40	37.5	242	– 2	– 484
40–45	42.5	220	– 1	– 220
45–50	47.5	268	0	0
50–55	52.5	70	+ 1	+ 70
			N = 800	∑fd = – 634

$$\overline{X} = A + \frac{\Sigma fX}{N} \times i = 47.5 + \frac{634}{800} \times 5 = 47.5 - 3.96 = 43.54 \text{ years.}$$

Example 16:

A machinery depreciated by 40% in the first year 25 percent in the second and 10% p.a. during the next three years. What is the average rate of depreciation during the whole period?

Solution:

The cost of the machine will not affect of the rate of depreciation and hence it can be ignored. Average rate of depreciation would be obtained by applying geometric mean:

Year	*Diminishing value taking 100 as base X*	*log X*
1994	100 – 40 = 60	1,7782
1995	100 – 25 = 75	1,8751
1996	100 – 10 = 90	1,9542
1997	100 – 10 = 90	1,9542
1998	100 – 10 = 90	1,9542
		Σ log X = 9,5159

$$\text{G.M.} = \text{A.L.}\left(\frac{\Sigma \log X}{N}\right) = \left(\frac{9{,}5159}{5}\right) = \text{A.L.}1{,}9032 = 80.$$

Since the diminishing value is Rs. 80, the depreciation will be 100 – 80 = 20%.

Thus, the average rate of depreciation charged during the whole period is 20 percent.

Example 17

In a class of 50 students 10 have failed and their average of marks is 2.5. The total marks secured by the entire class were 281. Find the average marks of those who have passed.

Solution:

N = 50, No. of failures = 10, average marks of those who failed = 2.5.

Total marks secured by all 50 students = 281

Marks secured by 10 students who failed = 10 × 2.5 = 25

Marks secured by 50 students = 281

Out of 50 ten have failed.

Total marks of those who passed = 281 – 25 = 256

$$\text{Average marks of those who passed} = \frac{256}{400} = 6.4.$$

Example 18:

Draw the "less than" ogive for the data given below and answer the following from the graph:

Marks	*No. of candidates*	*Marks*	*No. of candidates*
0–10	*5*	*50–60*	*65*
10–20	*20*	*60–70*	*50*
20–30	*40*	*70–80*	*35*
30–40	*70*	*80–90*	*20*
40–50	*85*	*90–100*	*10*

(i) Determine the median and the two quartiles.

(ii) If the pass mark is 40, what percentage of candidates pass the examination?

Solution:

Drawing less than ogive

Marks	*f*	*Marks*	*f*
less than 10	5	less than 60	285
less than 20	25	less than 70	335
less than 30	65	less than 80	370
less than 40	135	less than 90	390
less than 50	220	less than 100	400

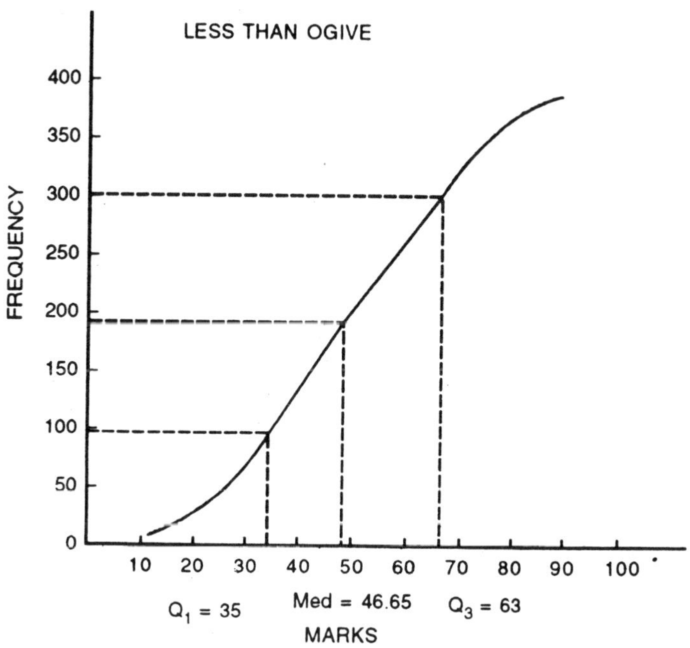

(i) Median = size of $\frac{N}{2}$ th item = $\frac{400}{2}$ = 20th item

Looking to the graph, median = 47.65

Q_1 = Size of $\frac{N}{2}$ item = $\frac{400}{4}$ = 100th item

Size of 100th item from the graph is = 35

Q_3 = Size of $\frac{3N}{2}$ = $\frac{400 \times 3}{4}$ = 300th item. Hence Q_3 = 63

(ii) Percentage of candidates passing exam.

Total no. of candidates getting 40 marks and above

= 400 – 135 = 265

Percentage of candidates passing

$$\frac{256}{400} \times 100 = 65.$$

Example 19:

The median and mode of the following wage distribution are known to be Rs. 33.5 and 34 respectively. Three frequency values from the table are however missing. Find these missing values.

Wages in Rs.	***Frequencies***
0–10	*4*
10–20	*16*
20–30	*?*
30–40	*?*
40–50	*?*
50–60	*6*
60–70	*4*
	230

Solution:

Let the missing frequencies be:

20–30	x
30–40	y
40–50	230–30–x–y

Since median and mode are 33.5 and 34 they both lie in the class 30–40.

$$\text{Med.} = L + \frac{N/2 - c.f.}{f} \times i$$

$$\text{Med.} = 33.5,\ L = 30,\ N/2 = 115,$$

$$c.f. = 20 + x,\ f = y,\ i = 10$$

$$33.5 = 30 + \frac{115 - (20 + x)}{y} \times 10$$

$$3.5 = \frac{95 - x}{y} \times \text{ or } 3.5y = 950 - 10x$$

$$\Rightarrow \qquad 3.5y + 10x = 950 \qquad \ldots(i)$$

$$\text{Mode} = L + \frac{\Delta_1}{\Delta_1 + \Delta_2} \times i$$

$$\text{Mode} = 34 = 30;$$

$$\Delta_1 = (y - x);$$

$$\Delta_2 = [y - (230 - 30 - x - y)];\ i = 10$$

$$34 = 30 + \frac{y - x}{y - x + 2y - 200 + x} \times 10$$

$$= 30 + \frac{(y - x)}{3y - 200} \times 10$$

$$4 = \frac{10y - 10x}{3y - 200} \text{ or } 12y - 800 = 10y - 10x$$

$$\Rightarrow \qquad 2y + 10x = 800 \qquad \ldots(ii)$$

Multiplying Eqn. (i) by 10

$$3.5y + 10x = 950$$

$$2y + 10x = 800$$

$$1.5y = 150$$

$$y = 100$$

Substituting the value of y in Eqn. (i)

$$3.5 \times 100 + 10x = 950$$

$$\Rightarrow \qquad 350 + 10x = 950 \text{ or } x = 960$$

Thus the missing frequencies are:

Class intervals	*Frequencies*
20–30	x = 60
30–40	y = 100
40–50	200 – 60 – 100 = 40

Example 20:

An economy grows at the rate of 2% in the first year, 2.5% in the second year, 3% in the third year, 4% in the fourth year ... and 10% in the tenth year. What is the average rate of growth of the company?

Solution:

For finding the average rate of growth we shall have to calculate the geometric mean. Since the growth rate is 4% in the fourth year and 10% in the 10th year and in between the growth rates are not given we shall assume that the growth rates have been 5 6 7% etc. in the 5th, 6th, 7th year respectively.

Year	*Growth Rate*	*Value at the end of the year x*	*log x*
1st	2	102	2.0086
2nd	2.5	102.5	2.0107
3rd	3	103	2.0128
4th	4	104	2.0170
5th	5	105	2.0212
6th	6	106	2.0253
7th	7	107	2.0294
8th	8	108	2.0334
9th	9	109	2.0374
10th	10	110	2.0414
			$\Sigma \log X = 20.2372$

$$\text{G.M.} = \text{AL}\left(\frac{\Sigma \log X}{N} = \text{AL}\left(\frac{20.2372}{10}\right)\right) = \text{AL}(2.022372) = 105.6$$

Average rate of growth = 105.6 – 100 = 5.6 percent.

Example 21:

Incomes of employees in an industrial concern are given below. The total income of the 10 employees in the class over Rs. 2,500 is Rs. 30,000. Compute the mean income. Every employee belonging to the top 25% of the earners is required to pay 5% of 'income to workers' relief fund. Estimate the contribution to this fund.

Income (Rs.)	*Frequency*	*Income (Rs.)*	*Frequency*
Below 500	*90*	*1500–2000*	*60*
500–1000	*150*	*2000–2500*	*70*
1000–1500	*100*	*2500 and cover*	*10*

Solution:

Compertation of Arithmetic Mean

Income (Rs.)	*M.P. m*	*f*	*fm*
0–500	250	90	22500
500–1000	750	150	112500
1000–1500	1250	100	125000
1500–2000	1750	80	140000
2000–2500	2250	70	157500
2500 and over	—	10	30000
		N = 500	Σ fm = 587500

$$\overline{X}\ \frac{\Sigma fm}{N} = \frac{587500}{500} = 1175$$

Note: If is given that the total income of 10 employees in the class over Rs. 2,500 is Rs. 30,000.

Number of employees belonging to the top 25% of the earners is

$$\frac{25}{100} \times 500 = 125$$

and the distribution of these top earners (as obvious from the above table) is as follows:

Distribution of Top 25% of Earners

Income (Rs.)	*Frequency*
2500 and over	10
2000–2500	70
1500–2000	45

45 persons in the last class *i.e.,* 1500–2000 are to be taken with the highest income level starting from 2000 and below. Under the assumption

that the frequencies are distributed uniformly over the entire class we interpolate this number as follow:

80 persons have income in the range 2000 – 1500 = Rs. 500

∴ 45 persons have income in the range

$$\frac{500}{60} \times 45 = 281.25 \text{ or } 281.$$

Since we are interested in the top 45 earners in the income group 1500 – 2600 their salaries will range from (2000 – 281) to 2000 *i.e.* 1719 to 2000.

The distribution of top 125 persons is as follows:

Income (Rs.)	*Mid-point m*	*f*	*Total Income fm*
2500 and over	—	10	30,000 (given)
2000–2500	2250	70	1,57,500
1719–2000	1859.5	45	83,677.5
		N = 125	2,71,177.5

Hence the total income of the top 25% of earners is Rs. 2,71,177.5.

5% contribution to the fund

0.05 of 2,71,177.5 = Rs. 93,598.

Example 22:

(a) A man travelled by cr for 3 days. He covered 480 km. each day. On the first day he drove for 10 hour on the second day he drove for 12 hours at 40 km. an hour and on last day he drove for 15 hours at 32 km. What was his average speed?

Solution:

Since the distance travelled is constant *i.e.* 480 km. each day the appropriate erage is the Harmonic Mean.

$$\text{H.M.} = \frac{N}{\frac{1}{X_1} + \frac{1}{X_2} + \frac{1}{X_3}}$$

$$= \frac{3}{\frac{1}{48} + \frac{1}{40} + \frac{1}{32}} = \frac{3}{\frac{37}{480}} = \frac{3 \times 480}{37} = 38.92 \text{ km p.h}$$

Mr. Dushmanta of Bhubaneswar started for village which was at a distance of six km. travelled in his car at a speed of 40 km. per hour. After travelling for 4 km. the car topped running. He then travelled in a rickshaw at a speed of 10 km. per hour After travelling a distance of $1\frac{1}{2}$ km. he left rickshaw and covered the remaining distance on foot a speed of 4 km. per hour.

Find the average speed per hour of Mr. Dushmanta and verify the calculations.

Solution:

This problem can be solved with the help of Weighted harmonic Mean.

Speed X	*Distance W*	*W/X*
40	4.0	0.100
10	1.5	0.150
4	0.5	0.125
	ΣW = 6	**Σ(W/X) = 0.375**

$$\text{Average speed} = \frac{\sum W}{\sum (W/X)} \ \frac{6}{0.375} = 16 \text{ km. p.h.}$$

Example 23:

In a factory there are 100 skilled, 250 semi-skilled and 150 unskilled workers. It has be observed that on an average a unit length of a particular fabric is woven by a skilled worker in 3 hours by a semi-skilled worker in 4 hours and by an unskilled worker in 5 hours. After a training of 2 years the semiskilled workers are expected to become skilled and the training for weaving the unit length of fabric by an average worker?

Solution:

Average time per worker before training

$$\frac{(100 \times 3) + (250 \times 4) + (150 \times 5)}{100 + 250 + 150} = \frac{2050}{500} = 4.1 \text{ hrs.}$$

Now after training the composition of workers is as follows :

Skilled workers = 100 + 250 = 350

Semi-skilled workers = 150

Unskilled workers = Nil

Average time per worker after training is :

$$\frac{(350 \times 3) + (150 \times 4)}{350 + 150} = \frac{1050 + 600}{500} = 3.3 \text{ hours.}$$

After 2 years one hour less would be required.

Note. An assumption has even been made that there has not been any turnover of workers.

Example 24:

Calculate the median and mod (by grouping method) form the following data:

Central size:	*15*	*25*	*35*	*45*	*55*	*65*	*75*	*85*
Frequencies:	*5*	*9*	*13*	*21*	*20*	*15*	*8*	*3*

Solution:

Since we are given central values first we determine the lower and upper limits of the classes. The class interval is 10 and hence the first class would be 10-20 (because the mid-point is 15)

Calculation of Median and Mode

Class groups	*m.p.m*	*f*	*(m – 55)/10 d*	*fd*	*c.f.*
10-20	15	5	–4	–20	5
20-30	25	9	–3	–27	14
30-40	35	13	–2	–26	27
40-50	45	21	–1	–21	48
30-60	55	20	0	0	68
60-70	65	15	+1	+15	83
70-80	75	8	+2	+16	91
80-90	85	3	+3	+9	94
		N = 94		Σfd = 54	

Calculation of Median: Med. = size of $\frac{N}{2}$ th item = $\frac{94}{2}$ = 47th item

Median lies in the class 40-50.

$$\text{Med.} = \frac{N/2 - c.f.}{f} \times i\ 40 + \frac{47 - 27}{21} \times 10 = 40 + 9.524 = 49.524$$

Class groups	*I*	*II*	*III*	*IV*	*V*	*VI*
10-20	5					
		14		27		
20-30	9		22		43	
30-40	13	34				
40-50	21		41			54
50-60	20			56		
		35			43	
60-70	15		23			
70-80	8					26
		11				
80-90	3					

Class in which Mode is Expected to lie

Col. No.	*40-50*	*50-60*	*60-70*
I	1		
II		1	1
III	1	1	
IV	1	1	1
V	1	1	
VI	1	1	1
	5	5	3

This series is bimodal and hence we shall determine the mode by the indirect method.

Mode = 3 Median – 2 Mean.

Calculation of Mean $\overline{X} = A + \frac{\Sigma fd}{N} \times i = 55 - \frac{54}{94} \times 10$

$= 55 - 5.745 - 49.255$

Mode = (49.524) – 2 (49.255) = 148.572 – 98.51 = 50.062

Example 25:

Find the missing frequency from the following data :

Marks	:	*0 – 10*	*10 – 20*	*20 – 30*	*30 – 40*	*40 – 50*	*50 – 60*
No. of students	:	*5*	*15*	*20*	*–*	*20*	*10*

The arithmetic mean in 34 marks.

Solution:

Let the missing frequency be denoted by X.

Calculation of Missing Frequency

Marks	*f*	*m.p.*	*fm*
0-10	5	5	25
10-20	15	15	225
20-30	20	25	500
30-40	X	35	35*
40-50	20	45	900
50-60	10	55	550
		N = 70 + X	Σfm = 2200 + 35 X

$$\overline{X} = \frac{\Sigma fX}{N},\ 34 = \frac{22000 + 35X}{70 + 35X}$$

$$34(70\ '\ X) = 2200 + 35X$$

$$2380 + 34X = 2200 + 35X$$

$$34X - 35X = 2200 - 2380$$

$-X = 180$ or $X = 180$ Hence the missing value is 180.

Note. By putting X = 180 we can calculate the arithmetic mean and verify the result. It comes out to be the same *i.e.* arithmetic mean is 34.

Example 26:

In a class of 50 students 10 have filed and their average of marks is 2.5. The total marks secured by the entire class were 28.1. Find the average marks of hose who have passed.

Solution:

N = 50 failed = 10

Mean marks of those who failed = 2.5

Total marks of 10 students who failed = 2.5 × 10 = 10 = 25

Total marks secured by entire class = 281

Total marks obtained by those who passed = 281 – 25 = 256

Average marks obtained by those who passed = $\frac{256}{40}$ = 6.4

Example 27:

Average rainfall of a city from Monday to Saturday is 0.3 inch. Due to heavy rainfall on Sunday the average rainfall for the week increased to 0.5 inch. What was the rainfall on Sunday?

Solution:

Average rainfall from Monday to Saturday = 0.3 (*i.e.* for six days).

Total rainfall for the entire week including Sunday = 0.5

∴ Total rainfall = 5 × 7 = 3.5

Hence rainfall on Sunday = 3.5 – 1.8 = 1.7.

Example 28:

From the following data calculate the missing value when mean is 115.86:

Wages (Rs.)	:	*110*	*112*	*113*	*117*	*(X)*	*125*
No. of workers	:	*25*	*17*	*13*	*15*	*14*	*08*

Solution:

Calculation of Missing Value

Wages (Rs.)	*f*	*fX*
110	25	2750
112	17	1904
113	13	1469
117	15	1755
X	14	14X
125	8	1000
128	6	768
130	2	260
	N = 100	**ΣfX = 9906 = 14X**

$$\overline{X}\ \frac{\Sigma fd}{N} \Rightarrow 115.86 = \frac{9906 + 14X}{100}$$

11586 = 9906 + 14X 14X = 1680

$\Rightarrow X = 120$

Hence the missing value is 12.0

Example 29:

The rate of a certain commodity in the first week of January 1987 is 0.4 kg. per rupee : it is 0.6 kg per rupee in the third week. Therefore is it correct to say that the average price is 0.5 kg. per rupee? Verify.

Solution:

The answer given is not correct as it is based on arithmetic mean whereas in this base appropriate average is harmonic mean.

$$H.M. = \frac{N}{\frac{1}{a} + \frac{1}{b} + \frac{1}{c}}$$

$$= \frac{3}{\frac{1}{0.4} + \frac{1}{0.6} + \frac{1}{0.5}} = \frac{.3 + .2 + .24}{12} = \frac{3}{\frac{.74}{.12}} = \frac{3 \times .12}{.74} = 0.486$$

Hence the average price is 0.486 kg. rupee and not 0.5 kg. per rupee.

Example 30:

Find the missing frequency. If arithmetic mean is 28 of the data given below find the median of the series later.

Profits per shop :	*0 – 10*	*10 – 20*	*20 – 30*	*30 – 40*	*40 – 50*	*50 – 60*
No. of shops :	*12*	*18*	*27*	–	*17*	*6*

Solution:

Let the missing frequency by X.

Determination of Messing Frequency

Profits (Rs.)	*m.p.m*	*No. of shops f*	*(m – 35)/10 d*	*fd.*
0-10	5	12	–3	–36
10-20	15	18	–2	–36
20-30	25	27	–1	–27
30-40	35	X	0	0
40-50	45	17	+1	+17
50-60	55	6	+2	+12
		N = 80 + X		Σfd = – 70

$$\overline{X} = A + \frac{\Sigma fd}{N} \times i$$

$A = 35\ \Sigma fd = -70\ N = 80 + Xi = 10\ \overline{X} = 28$ (given).

$$28 = 35 \frac{70}{80 \times X} \times 10$$

$28(80 + X) = 35(80 + X) - 700$

$2240 + 28X = 2800 + 35X - 700$

$28X - 35X = 2800 - 700 - 2240$

$-7x = -140$ or $X = 20$.

Hence the missing frequency is 20.

Calculation of Median

Profits (Rs.)	*No. of shops* *f*	*c.f*
0-10	12	12
10-20	18	30
20-30	27	57
30-40	20	77
40-50	17	94
50-60	6	100
		N = 100

Med. size of $\frac{N}{2}$th item = $\frac{100}{2}$ = 50th item

Median lies in the class 20-30.

$$\text{Med} = L + \frac{N/2 - c.f.}{f} \times i$$

$L = 20\ N/2 = 50\ c.f. = 30\ f = 27\ i = 10$

$$\text{Med.} = 20 \frac{50 - 30}{27} \times 10 = 20 + 7.407 = 27.\ 407.$$

Example 31:

Calculate Median and Arithmetic Mean from the following series:

05 men get less than Rs. 5

12 men get less than Rs. 10

22 men get less than Rs. 15

30 men get less than Rs. 20

36 men get less than Rs. 25

40 men get less than Rs. 30 .

Solution:

We are give cumulative frequencies. First find simple frequency and then compute arithmetic mean and median.

Income (Rs.)	*m.p.m*	*f*	*(m-12.5)/5d*	*fd*	*c.f.*
0-5	2.5	5	–2	–10	5
5-10	7.5	7	–1	–7	12
10-15	12.5	10	0	–0	22
15-20	17.5	8	+1	+8	30
20-25	22.5	6	+2	+12	36
25-30	27.5	4	+3	+12	40
			N = 40		Σfd = 15

$$\overline{X} = A + \frac{\Sigma fd}{N} \times 12.5 + \frac{15}{40} \times 5\ 12.5 + 1.875 = 14.375$$

$$\text{Med.} = \text{Size of } \frac{N}{2}\text{th item} = \frac{40}{2} = 20\text{th item}$$

Median lies in the class 10-15.

$$\text{Med} = L + \frac{N/2 - c.f.}{f} \times i$$

$$= 10 + \frac{20 - 12}{10} \times 5 = 10 + 4 = 14$$

Example 32:

Find the missing information in the following table:

	A	*B*	*C*	*Combined*
Number	*10*	*8*	–	*24*
Mean	*20*	–	*6*	*15*
geometric Mean	*10*	*7*	–	*8.397*

Solution:

Finding missing information:

Number: For c missing information shall be

$$24 - (10 + 8) = 6$$

Mean: Let x be the mean of B

Them $(20 \times 10) + (8 \times x) + (6 \times 6) = (15 \times 24)$

$$200 = 8x + 36 = 360$$

$$8x = 360 - 200 - 36 = 124$$

$$x = \frac{124}{8} = 15.5$$

Hence mean of $B = 15.5$

Geometric Mean: Let x be the geometric mean of C.

$$(10)^{10} \times (7)^8 \times x^6 = (8.397)^{24}$$

$$= 10 \log 10 + 8 \log 7 + 6 \log x = 24 \log 8.397$$

$$= 10 + (8 \times 0.8451) + 6 \log x = 24\ (.9241)$$

$$= 10 + 6.7608 + 6 \log x = 22.1784$$

$$6 \log x = 22.1784 + 10 - 6.7608 = 5.44176$$

$$\log x = 0.9029$$

$$x = \text{A.L. } .9029 = 7.997$$

Hence geometric mean of C is 7.997.

Example 33:

During a period of decline in stock market prices a stock sold at Rs. 50 per share on one day Rs. 40 on the next day and Rs 25 on the third.

(i) If an investor bought 100, 120 and 180 shares on the respective three days, find the average price paid per share.

(ii) If the investor bought Rs 1000 worth of shares on each of the three days find the average price paid per share.

Solution: (i)

x	*w*	*wx*
50	100	5000
40	120	4800
25	180	4500
	Σw − 400	**Σwx = 14300**

$$\text{Average price paid per share} = \frac{\Sigma WX}{\Sigma W} = \frac{14300}{400} = 35.75$$

Hence average price per share = Rs. 35.75

(ii) The investor buys shares of Rs. 1,000 per day. We can calculate the number of shares bought on each day

Price	*Amount spent per day*	*No. of shares bought*
50	1000	20
40	1000	25
25	1000	40
		105

Hence average price paid per share $\frac{105}{3}$ = Rs. 35.

EXERCISES

1. (a) Calculate mean, median and mode from the following frequency distribution of marks at a test in English.

Marks :	5	10	15	20	25	30	35	40	45	50
No. of students :	20	43	75	76	72	45	39	9	8	6

(b) Geometric mean of 2 numbers is 15. If by mistake one figure is taken as 5, instead of 3, find correct geometric mean.

2. (a) A market with 330 firms has the following distribution of average number of laborers in different income groups:

Incommensurable *(value) Rs.* :	450	750	1050	1350	1650
No. of firms :	80	64	52	56	78
Average no. of workers :	16	24	15	17	8

Find the median income of all the orders.

(b) Determine the Geometric mean for the following values:

X : 5 8 12 16

3. (a) How Would you account for the predominant choice of arithmetic mean as a measure of central tendency? Under what circumstances would it be appropriate to use mode of median?

(b) Ram buys 1 kg. oranges from each of the four places at the rate of 2kg 4kg., 5kg., 8 kg., per rupee respectively. Find out how many oranges per rupee he purchased on an average?

4. The following is frequency distribution of the diameter of 1,000 parts of a particular type produced by the ABC Co. Ltd. Find the mean diameter of these parts.

Diameter (in centimeters)	*Number of parts*	*Diameter (in centimetres)*	*Number of parts*
3.05 to 3.14	13	3.55 to 3.64	296
3.15 to 3.24	29	3.65 to 3.74	74
3.25 to 3.34	65	3.75 to 3.84	40
3.35 to 3.44	200	3.85 to 3.94	5
3.45 to 3.54	278		

5. Following,distribution gives the pattern of overtime work done by 100 employees of a company. Calculate median, first quartile, and 7th decile.

Overtime hours :	10–15	15–20	20–25	25–30	30–35	35–40
No. of employees :	11	20	35	20	8	6

6. The monthly salary distribution of 250 families in a certain locality in Agra is given below:

Monthly Salary (Rs.)	*No. of Families*	*Monthly Salary (Rs.)*	*No. of Families*
More than 0	250	More than 2,000	55
More than 500	200	More than 2,500	30
More than 1,000	120	More than 3,000	15
More than 1,500	180	More than 3,500	5

Draw a 'less than' ogive for the data given above and hence find out:

(i) Limits of the income of middle 50% of the families: and

(ii) If income-tax is to be levied on families whose income exceeds Rs. 1,800 p.m., calculate the percentage of families, which will be paying income tax.

7. Calculate mean, median and mode from the following data:

Marks more than :	0	20	40	60	80	100	120
No. of students :	80	76	50	28	18	9	3

8. (a) Calculate mean, median, mode from the following data of the heights in inches of a group of students:

61, 62, 63 61, 63, 64, 60, 65, 63, 64, 66, 64,

Now suppose that a group of students, whose heights are 60, 96, 59, 68,67 and 70 inches, is added to the original group, find the median and mode of the combined group.

(b) From the following frequency distribution, find the class frequencies that are missing:

Intelligence Quotient	*No. of students*	*Intelligence Quotient*	*No. of students*
55-64	2	105-114	?
65-74	19	115-124	92
75-84	78	125-134	14
85-94	?	135-134	4
95-104	301		

You are given that the total frequency is 900 and the median 100.48.

9 (a) What is statistical average? What are the desirable properties for an average to possess? Which of the average you know possess most of these properties?

(b) What is meant by measure of central tendency? What are the characteristics of a good measure of central tendency?

10. Find out the median and mode from the following table:

No. of days absent	*No. of students*	*No. of days absent*	*No. of students*
Less than 5	29	Less than 30	644
Less than 10	224	Less than 35	650
Less than 15	465	Less than 40	653
Less than 20	582	Less than 45	655
Less than 25	634		

11. A laundry uses two different makes of washing machines. According to its past experience, the following results have been recorded:

Make of Machine	*Median life*	*Mean Life*
A	6,500 hours	6,000 hours
B	6,000 hours	6,500 hours

If both makes are of the same price, which make should the laundry purchase from now on? Give reasons.

12. A travelling salesman made five trips during the past two months, making the sales as tabulated below:

Trip	*No. of days*	*Value of sales (Rs.)*	*Sales per day (Rs.)*
1	4	500	125
2	5	380	76
3	8	576	72
4	3	108	63
5	4	260	65
	N = 24	1.824	374

(a) The sales manager criticised the salesman's performance as not very good since his mean sales are only Rs. 74.8 $\left(\frac{374}{5} = 74.8\right)$. The salesman replied that the sales manager was unfair in making such a statement, for his mean sales were as high as Rs. 76 $\left(\frac{1824}{24} = 76\right)$. What does each average mentioned here mean?

(b) Which mean seems to you appropriate in this case?

13. From the following data of the heights of 100 persons in a commercial concern, determine the modal height:

Height is inches :	58	60	61	62	63	64	65	66	68	70
No. of persons :	4	6	5	10	20	22	24	6	2	1

14. (a) The percentage of marks of 20 students, who were eligible for scholarships, are given below:

52	62	51	71	54	61	53	51	50	57
64	56	54	69	63	65	59	58	68	57

Calculate the monthly scholarship payable to the students as well as the average scholarship.

(b) Calculate mean, median and mode for the following frequency distribution:

Marks	***No. of students***	***Marks***	***No. of students***
0–20	21	51–60	24
21–30	19	61–70	18
31–40	60	71 80	15
41–50	42		

15. (a) The mean age of a group of 100 persons was found to be 32.02. Later, it was discovered that age 57 was misread as 27. Find correct mean.

(b) The goniometric mean of 10 observations was calculated as 28.6. It was later discovered that one of the observations was recorded as 23.4 instead of 32.4. Apply appropriate correction and calculate the correct geometric mean.

16. Compute Mean. Mcdian and Mode from the following data:

Income (Rs.): More than	10	20	30	40	50	60	70	80
No. of persons:	72	67	59	50	36	21	9	3

17. Calculate mean, median and mode from the following data:

Marks	*No. of students*	*Marks*	*No. of students*
10–20	4	10–60	124
10–30	16	10–70	137
10–40	56	10–80	146
10–50	95	10–90	150

18. (a) Calculate the weighted geometric mean of the following data:

Weights (in lbs):	130	135	140	145	146	148	149	150	157
No. of Workers:	3	4	6	6	—	3	5	2	1

(b) Find out the missing frequencies for the class intervals using following data on 150 students:

Marks	*No. of students*	*Marks*	*No. of students*
0–10	4	50–60	25
10–20	8	60–70	40
20–30	11	70–80	?
30–40	15	80–90	2
40–50	?	90–100	1

Given that the mean marks are 65.

19 (a) Prove that the arithmetic mean of two positive numbers a and b is at least as large as their geometric mean.

(b) Find the missing figure:

Median = Mode + ? (Mean–Mode).

20. (a) What are the functions of an average?

(b) Explain the method of drawing a Lorenz curve with the help of an example.

21. (a) What characteristics a good average should possess? How far does the median meet these requirements? Explain the concept and uses of weighted average.

(b) "Every average has its own peculiar characteristics. It is difficult to say which average is the best". Comment briefly.

22. (a) Why is arithmetic mean considered to be the most suitable measure of central tendency?

(b) State the empirical relationship between mean, median, and mode.

23. (a) Discuss the merits and demerits of median as a measure of central tendency. Also calculate the two quartiles Q_1 and Q_3 from the following data.

    (b) Describe the different Measure of central tendency of a frequency distribution, mentioning their merits and demerits.

    (c) Briefly explain the role of Grouping Table and Analysis Table in calculation of Mode.

24. (a) What are the properties of a good average? Examine these properties with reference to the arithmetic mean and geometric mean.

    (b) Point out the difference between the simple arithmetic mean and weighted arithmetic mean.

    (c) A man travels 20 kms at 40 kms per hours, 10 kms at 60 kms per hours. What is his average speed?

25. (a) What do your understand by central tendency? Explain with the help of an example. What purpose does a measure of central tendency serve?

    (b) Explain clearly the concept of central value taking a suitable example. Does central value always imply middle most value?

26. (a) What are the various Measure of location of distribution and for what purposes are they used?

    (b) The purpose of an average is to represent a group of individual values in simple and concise manner so that the mind can get a quick understanding of general size of individuals in the group. Explain.

    (c) State the various Measure of central tendency.

    (d) An average is a substitute for complex group of variables but its is not always safe to depend on the substitute all to the exclusion of individual members of the group Discuss.

27. (a) A motor car covered a distance of 50 miles four times. The first time at 50 m.p.h., the second at 20 m.p.h., the third at 40 m.p.h. and the fourth at 25 m.p.h. Calculate the average speed and explain the choice of the average.

    (b) A man gets three annual raises in salary. At the end of the first year, he gets an increase of 4%, at the end of second. An increase of 6% on his salary as it was at the end of the year, and at the end of the third year, an increase of 9% on his salary as it was at the end of the third year. What is the average percentage increase?

(c) A machine is assumed to depreciate 40 per enc in value in the first year, 25% in the second year and 10% per annum for the next three years, each percentage being calculated on the diminishing value. What is the average percentage depreciation for five years?

28. (a) Mr. A spends Rs. 100 for apples costing Rs. 10 per kilogram and another Rs. 100 for apples costing Rs. 8 kilogram. What is the average price per kilogram?

(b) A company expects the sale of its products to in crease by 50% in the following year. By what percentage should it increase the selling price in order to double its gross revenue?

(c) Three men take 12, 8, 6 hours respectively to husk an acre of corn. Determine the average number of hours to husk an acre.

(d) In a moderately skewed distribution arithmetic mean = 24.6 and the mode = 16.1. Find the value of the median and explain the reason for the method employed.

29. (a) The price of a commodity increased by 20% from 1997 to 1988, by 15% from 1998 to 1999 and by 10% from 1999 to 2000. Find the yearly average percentage increase from 1997 to 2000.

(b) Calculate geometric mean and harmonic mean of the following data:

125, 1462, 3, 7, 0.22, 0.08, 12.75, 0.5

30. Classify the following data of 50 marks obtained in statistics subject appropriately and compute the arithmetic and geometric means:

64,	26,	56,	85,	42,	38,	10,	55,	72,	65,
15,	47,	59,	62,	52,	54,	48,	62,	81,	31,
44,	54,	50,	52,	66,	38,	77,	88,	17,	58,
15,	25,	49,	51,	64,	68,	53,	50,	72,	58,
50,	61,	70,	80,	90,	05,	16,	51,	61,	52,

31. How would you account for the predominant choice of arithmetic mean of statistical data as a measure of central tendency? Under what circumstances would it be appropriate to use mean. mode and median? Discuss.

32. (a) Compare and contrast arithmetic mean. Geometric mean and harmonic mean. Which of them is least affected by extreme items?

(b) What are the desirable properties of a good average? Under what circumstances the use of the median harmonic and geometric means would give better results than the arithmetic mean? Explain with examples.

33. What do you mean by 'Central Tendency'? What are the desirable properties for an average? Which average possesses most of these properties?

34. (a) Discuss the merits and demerits of geometric mean. Explain its utility and algebraic characteristics.

    (b) Indicate briefly which of the properties of good measure of central tendency are possessed by any two of the following:

    Arithmetic mean, median, mode, geometric mean and harmonic mean.

35. (a) Explain what is meant by weighted average and discuss the effect of weighting.

    (b) Under what circumstances geometric mean and harmonic mean are suitable Measure for describing the central tendency of a frequency distribution?

    (c) What is an average? Explain the geometric mean and state its merits and limitations.

36. Construct a frequency table for the following data regarding annual profits (in thousands of rupees) of 50 firms, taking 25–34, 35–44, etc. a class intervals:

28	35	61	29	36	48	57	67	69	50
48	40	47	42	41	37	51	62	63	33
31	32	35	40	38	37	60	51	54	56
37	46	42	38	61	59	58	44	39	57
38	44	45	45	47	38	44	47	47	64

Construct a less then ogive and find:

(a) Number of firms having profit between 37.000 and Rs. 50,000

(b) Middle 50% group's range.

37. Calculate mean, median, media, mode for the following data:

Weekly *Earnings (Rs)* :	66-67	6768	68-69	69-70	70-71	71-22
No. of Persons :	15	24	40	20	14	11

38. (a) Draw a less than ogive from the following and locate the median:

Size :	10-20	20-30	30-40	40-50	50-60
Frequency :	20	60	100	150	75

(b) Calculate weighted arithmetic mean from the following data:

X :	1500	800	500	250	100
W :	10	20	70	100	150

39. The following table shows the distribution of families according to their daily expenditure. The median and mode for distribution are Rs. 24.8 and Rs. 24, respectively. Find missing frequencies and A.M. ($\overline{X}$) of data.

Expenditure (Rs.) :	0-10	10-20	20-30	30-40	40-50
No. of Families :	14	–	27	–	15

40. An incomplete distribution is given as follows:

Variable :	0-10	10-20	20-30	30-40	40-50	50-60	60-70
Frequency :	10	20	?	40	?	25	15

You are given that the median value is 35. Using the median formula, fill up the missing frequencies.

41. From a batch of 13 students, who had appeared for an examination. 4 students have failed. The marks of the successful candidates were 41, 57, 38, 61, 36, 35, 71, 50, 40, Calculate the median marks.

42. The following table gives the weekly wages in rupees in a certain commercial organisation

Weekly wages (Rs.) :	30-32	32-34	34-36	36-38	38-40	40-42
Frequency :	2	9	25	30	49	62
Weekly wages (Rs.) :	42-44	44-46	46-48	48-50		
Frequency :	39	20	99	3		

Calculate from the above data:

(i) the median and the third quartile wages: and

(ii) the number of wage emmers receiving between Rs. 37 and Rs. 46 per week.

43. (a) Give a brief note of the Measure of central tendency together with their merits and demerits. Which is the best measure of central tendency and why?

(b) Under what circumstances would it be appropriate to use arithmetic mean, median and mode? Discuss.

44. (a) "The arithmetic mean is the best among all the averages." Given reasons to substantiate the statement. State the purpose of studying the other averages.

(b) Compare and contrast various Measure of central tendency.

45. (a) An investor purchased securities of a company investing a sum of Rs. 2,400 every month. If he bought these securities at Rs. 15 during the three months respectively what is the average price paid by him?

(b) Calculate the mean, median and mode for the following frequency distribution and verify the empirical relation connecting them:

Marks :	0-20	20-40	40-60	60-80	80-100
Frequency :	3	17	27	20	9

46. (a) From the information given below, find:

(i) Which factory pays larger amount as daily wages?

(ii) What is the average daily wage for the workers of two factories?

	Factory A	***Factory B***
No. of wage earners	250	200
Average daily wages	Rs. 12.0	Rs. 13.8

(b) A cyclist covers his first three kinds at an average speed of 8 km. Per hour, another two kms at 9 kms. per hour and the last two km. at 2 kms. per hour, Find the average speed for the entire journey.

(c) The average weight of a group of 25 boys was calculated to be 78.4 lb. It was later discovered that weight of one boy was misread as 69 lb. instead of the correct value 96 lb. Calculate the correct average.

(a) $\overline{X}_{12}$ = 12.8, (b) 4.3 km., (c) 79.5

47. Give a specific example of your own for each of the following cases:

(a) The Indian is preferred to the arithmetic mean.

(b) The geometric mean would be more satisfactory than the arithmetic mean.

(c) The median would be preferred to the mode.

(d) The mode would be preferred to the median.

(e) The harmonic mean must be used instead of the arithmetic mean.

(f) No average would be meaningful.

48. (a) Draw less than cumulative frequency curve for the following distribution. Read the median from the graph and verify your result by the mathematical formula. Also obtain the limits of monthly income of central 50% of the employees:

Monthly Income (Rs.)	*No. of Employees*
Below 2000	3
2000-2200	7
2200-2400	25
2400-2600	30
2600-2800	24
2800-3000	9
3000 and above	2

(b) Calculate mean and median for the following data:

Profits (Rs. Lakhs)	*Frequency*	*Profits (Rs. Lakhs)*	*Frequency*
10-20	4	10-60	124
10-30	16	10-70	137
10-40	56	10-80	146
10-50	97	10-90	150

(c) The mean age of a combined group of men and women is 25 years. If the mean age of the group of men is 26 and that of the group of women is 21. Find out the percentage of men and women in the group.

49. You are given below a distribution of income per month. Calculate the most suitable average giving reasons for your choice.

Income (Rs.)		*Frequency*
Less than 100	...	40
100-200	...	89
200-300	...	148
300-400	...	64
400 and above	...	39

50. Calculate the Median from the following data, if mean value is 44:

Marks	*No. of Students*	*Marks*	*No. of Students*
70-80	10	30-40	12
60-70	10	20-30	7
50-60	20	10-20	8
40-50	?	0-10	5

51. (a) What are averages? Calculate the mean and mode for the following frequency distribution:

Monthly wages (Rs.)	*No. of workers*	*Monthly wages (Rs.)*	*No. of workers*
Less than 200	78	600-800	42
200-400	165	800-1,000	12
400-600	93		

(b) Find out the quartiles and the mode for the following (take suitable class intervals):

Income (Rs.)	*No. of Persons*	*Income (Rs.)*	*No. of Persons*
Below 30	69	61-70	58
31-40	167	71-80	27
41-50	207	81 and above	10
51-60	65		

52. (a) A book seller has 150 books of Economics and Accountancy. The average price on these books is Rs. per book. Average price of books on Economics is Rs. 43 and that of Accountancy is Rs. 35. Find out the number of books on Economics with the seller?

(b) Draw ogive by less than and more than methods for the following weekly income distribution:

Weekly Income (Rs.)	*No. of employees*	*Weekly Income (Rs.)*	*No. of employees*
Below 550	5	700-750	16
550-600	10	750-800	12
600-650	22	Above 800	15
650-700	30		

Read the value of median from the graph and verify your value from the formula of median. Also obtain the limits of weekly income of central 50 percent of the employees.

53. (a) A man travelled from one city to another. The distance between the two cities is 4 kms. He drives his car at 40 kms per hour. After travelling one km., the car stops running. He then travels on a tonga at 10 km, per hour. After travelling a distance of 1.5 km, he leaves and covers the remaining distance on foot at 4 km, per hour. Find the average speed per hour of that person.

(b) (i) Draw an ogive curve form the following data and measure the median value. Verify it by actual calculations.

Central Size	:	5	15	25	35	45
Frequency	:	5	11	21	16	10

(ii) The mean weight of 150 workers in a factory is 56 kgs. If the mean weight of men in the factory is 64 kgs. and that of the women is 48 kg., find the number of men and women is the factory.

54. Draw an ogive curve (less than type) for the following data and hence find: (i) median: (ii) inter-quartile range:

Daily Wages (Rs.)	***No. of Workers***	***Daily Wages (Rs.)***	***No. of Workers***
1–5	7	21–25	24
6–10	10	26–30	18
11–15	15	31–35	10
16–20	30	36–40	6

55. (a) Define median. Indicate its merits and demerits.

(b) From the following data draw the Histogram and Ogive and determine the mode and the median graphically:

Marks	:	0–6	6–12	12–18	18–24	24–30	30–36
No. of students	:	4	8	15	20	12	6

(c) Calculate median and mode of the following series:

Size	:	6–10	11–15	16–20	21–25	26–30
Frequency	:	20	30	50	40	10

56. (i) Determine the value of mode with mean and median whose values are 20 and 22 respectively.

(ii) The arithmetic mean of two observations is 25 and their geometric mean is 15. Find their harmonic mean.

(iii) A cyclist covers this first five km. at an average speed of 10 km. p.h., another 3 km. at 8 km. p.h. and the last 2 km. p.h. Find the average speed for the entire journey.

57. An incomplete frequency distribution is given below:

Marks	:	0–10	10–20	20–30	30–40	40–50	50–60	60–70	Total
Frequency	:	4	16	–	–	–	6	3	230

Find the three missing frequencies of the table, given that median = 33.5 and mode = 34. Also calculate the mean using the empirical relation between mean. Median and mode.

58. (a) Obtain the value of median from the following data:

335, 384, 407, 672, 522, 777, 753, 2488, 1490.

(b) Calculate the mean from the following frequency table:

Mid points	*Frequency*	*Mid points*	*Frequency*
1	2	6	155
2	60	7	79
3	101	8	40
4	152	9	1
5	205		

59. (a) 'A' travelled some distance by cycle at a speed of 15 km. per hour. On return journey he travelled the same distance at a speed of 10 km per hour. What was his average speed per hour?

(b) The average monthly wage of all workers in a factory is Rs. 444. If the average wages paid to male and female workers are Rs. 480 and Rs. 360 respectively, find the percentage of male and female works employed by the factory.

60. (a) Determine the healthier town from the following information:

Age group (years)	*Town 'A' population*	*(Standard) Deaths*	*Town 'B' population*	*(Local) Deaths*
0–15	15,000	300	20,000	500
15–50	20,000	400	52,000	1,040
50 & above	5,000	140	8,000	240

(b) Calculate the value of mean for the following data:

Mid point :	10	15	20	25	30	35	40
Frequency :	7	9	18	26	10	4	3

61. (a) State, giving reasons, the average to be used in the following situations:

(i) To determine the average size of the shoe sold in the shop.

(ii) To determine the average wages in an industrial concern.

(iii) To find the averagê beauty among a group of students in a class.

(iv) To find the per capita income in different cities.

(b) State the formula of median for grouped data with class intervals.

(c) Under what circumstances is harmonic mean the most suitable?

(d) What is an average? Under what circumstances would you use the geometric mean instead of the arithmetic average?

(e) Explain the relationship between mean, median and mode in a symmetrical and moderately asymmetrical distribution.

62. (a) An aeroplane travels distances of d_1. d_2 and d_3 kms at speeds V_1, V_2 and V_3 kms per hour respectively. Show that the average speed is given by V_1 where:

$$\frac{d_1 + d_2 + d_3}{V} = \frac{d_1}{V_1} + \frac{d_2}{V_2} + \frac{d}{V}$$

(b) Calculate Geometric mean:

2 4, 8, 12, 16, 24.

63. Calculate the combined Arithmetic Mean of the following data:

	Class A	*Class B*
No. of Students	50	60
Average marks	64	59

64. What are the merits of the mode? Find out the mode from the following series:

Size :	0-5	5-10	10-15	15-20	20-25	25-30	30-35
Frequency :	1	2	5	14	10	9	2

65. (a) The following table gives the marks obtained by 60 students in statistics in certain examination. Find mean and median from the data:

Examination Marks	***No. of students***
More than 70%	7
More than 60%	18
More than 50%	40
More than 40%	40
More than 30%	63
More than 20%	65

(b) An income tax assessed depreciated the machinery of his factory by 20 percent in each of the first two years and 40 per cent in the third year. How much average depreciation relief should be claimed from taxation department?

(c) Compute the weighted mean of first natural numbers when weights are equal to the corresponding numbers.

66. The management of a college decides to give scholarship to the students who have scored marks 70 and above 70 in Business Statistics. The following are the marks scored by II B. Com. students:

71	73	74	85	86	88	91	94	96	99
74	74	76	93	91	94	96	98	88	94

The scholarship payable is given below:

Marks	*Scholarship amount (Rs.)*
70-75	100
75-80	200
80-85	300
85-90	400
90-95	500
95-100	600

Estimate the total scholarship payable and average scholarship

67. (a) Find the mean and mode from the following data:

% mark :	10-19	20-29	30-39	40-49	50-59	60-69	70-79
Students :	8	19	29	36	25	13	4

(b) Draw ogive curves for the following data and hence find the median:

Age (years)	*No. of persons*	*Age (years)*	*No. of persons*
20-25	9	40-45	19
25-30	25	45-50	13
30-35	34	50-55	7
35-40	25	55-60	2

68. Tick the correct answer from the following:

(a) The sum of the deviations of individual observations (i) mode, (ii) median, (iii) geometric mean, (iv) none of these.

(b) Which average is affected most by extreme observations (i) mode, (ii) median, (iii) geometric mean, (iv) arithmetic mean, (v) harmonic mean.

(c) In a moderately asymmetrical distribution:

(i) A.M. > G.M. > H.M.

(ii) A.M > G.M. > H.M.

(iii) A.M. > G.M. > H.M.

(iv) G.M > A.M. > H.M.

(v) H.M. > G.M. > A.M.

(d) Which of the following is the most unstable average: (i) mode, (ii) median, (iii) geometric mean, (iv) harmonic mean, (v) arithmetic mean.

(e) For dealing with qualitative data the best average is (i) geometric mean, (ii) median, (iii) harmonic mean. (iv) median, (v) mode.

(f) The positional measure of central tendency is (i) geometric mean, (ii) median, (iii) harmonic mean, (iv) arithmetic mean, (v) none of these.

(g) One of the methods of determining mode is (i) mode = 2 median –3 mean, (ii) mode = 2 median + 3 mean, (iii) mode = 2 median –2 mean. (iv) mode = 3 median + 2 mean, (v) none of these.

(h) The geometric mean of two numbers 8 and 18 shall be (i) 12. (ii) 13, (iii) 15, (iv) 11.08, (v) none of these.

(i) In a moderately skewed distribution the values of mean and median are 5 and 6 respectively. The value of mode in such a situation is approximately equal to:

(a) 8, (b) 11, (c) 16, (d) none of these

Ans. (a) (iv), (b) (iii), (c) (ii), (d) (i), (e) (iv), (f) (ii), (g) (iii), (h) (i).

69. Fill is the blanks:

(i) Median is better suited for ... interval series.

(ii) In a moderately asymmetrical distribution, the distance between the ...... and the ...... is about ...... the distance between the ...... and the ......

(iii) Given mean 25, mode 24, the median would be......

(iv) For studying phenomena like intelligence and honesty...... is a better average to used, while for phenomena like the size of shoes or redeemed garments. The average w be used is...

(v) In a symmetrical distribution mean... median.... mode.

(vi) The geometric mean is the....root of the product of all the measurements.

(vii) The mode a distribution is the value that has the greatest of....

(viii) The harmonic mean is the ...... of the arithmetic mean of the values.

Ans. (i) positional, (ii) mean, median, 1/3 mean, mode, (iii) 24.67, (iv) median, mode, (v) is equal to, is equal to (vi) nth, (vii) concentration, frequencies. (viii) reciprocal of.

4

MEASURES OF DISPERSION

List of Formulae

Individual Observations	**Discrete & Continuous Series**
Range Range = L = S Coeff. of Range = $\frac{L - S}{L + S}$	(Same as on the left) But L *i.e.*, largest value, will be upper limit of the highest class and S will be the lower limit of the lowest class
Quartile Devotion Q.D. = $\frac{Q_3 - Q_1}{2}$	(Same as on the left)
Coeff. of Q.D. = $\frac{Q_3 - Q_1}{Q_3 + Q_1}$	
Mean Deviation	
M. D. = $\frac{\sum \lvert D \rvert}{N}$	M.D. = $\frac{\sum f \lvert D \rvert}{N}$
Coeff. of M.D. = $\frac{M.D.}{Median}$	Coeff. of M.D. = $\frac{M.D.}{Median}$
or $\frac{M.D.}{Mean}$ (if deviations are taken from mean)	or $\frac{M.D.}{Mean}$ (if deviations are taken from mean)
Standard Deviation *Actual Mean Method* $\sigma = \sqrt{\frac{\left(\sum (X - X)^2\right)}{N}}$	*Actual Mean Method* $\sigma = \sqrt{\frac{\sum (X - \overline{X})^2}{N}}$

Assumed Mean Method	*Assumed Mean Method*
$\sigma = \sqrt{\frac{\sum d^2}{N} - \left(\frac{\sum d}{N}\right)^2}$	$\sigma = \sqrt{\frac{\sum fd^2}{N} - \left(\frac{\sum fd}{N}\right)^2}$
Step Deviation Method	*Step Deviation method*
$\sigma = \sqrt{\frac{\sum d^2}{N} - \left(\frac{\sum d}{N}\right)^2} \times i$	$\sigma = \sqrt{\frac{\sum fd^2}{N} - \left(\frac{\sum fd}{N}\right)^2} \times i$
$C.V. = \frac{\sigma}{X} \times 100$	$C.V. = \frac{\sigma}{X} \times 100$

Combined Standard Deviation

$$\sigma_{12} = \sqrt{\frac{N_1\sigma_1^2 + N_2\sigma_2^2 + N_1d_1^2 + N_2\sigma_2^2}{N_1 + N_2}}$$

where $d_1 = |\overline{X}_1 - \overline{X}_{12}|$ and $d_2 = |\overline{X}_2 - \overline{X}_{12}|$

INTRODUCTION

It is necessary to describe the variability or dispersion of the observations. In two or more distributions the central value may be the same but still there can be wide disparities in the formation of distribution. Measures of dispersion help us in studying this important characteristic of a distribution.

Most important definitions of dispersion are given below:

1. Dispersion or spread is the degree of the scatter or variation of the variable about a central value." —*Brooks & Dick*
2. "The measurement of the scatterness of the mass of figures in a series about an average is called measure of variation or dispersion". —*Simpson & Kafka*
3. "Dispersion is the measure of the variation of the items." —*A.L. Bowley*
4. "The degree to which numerical data tend to spread about an average value is called the variation of dispersion of the data." —*Spiegel*

Since measures of dispersion give an average of the differences of various items from an average, they are also called averages of the *second order.*

An average is more meaningful when it is examined in the light of dispersion.

The study of dispersion is of great significance in practice as could well be appreciated from the following example:

	Series A	*Series B*	*Series C*
	100	100	1
	100	105	489
	100	102	2
	100	103	3
	100	90	5
Total	500	500	500
$\overline{X}$	100	100	100

Since arithmetic mean is the same in all three series, one is likely to conclude that these series are alike in nature. But a close examination shall reveal that distributions differ widely form one another. In series A, each and every item is perfectly represented by the arithmetic mean or, in other words. None of the items of series A deviates from the arithmetic mean hence there is no dispersion. In series B, only one item is perfectly represented by the arithmetic mean and the other items vary but the variation is very small as compared to series C. In series C, not a single item is represented by the arithmetic mean and the items vary widely from one another. In series C dispersion is much greater compared to series B. Similarly, we may have two groups of laborers with the same mean salary and yet their distributions may differ widely.

The two curves in diagram (a) represent two distributions with the same mean $\overline{X}$, but with different dispersions. The two curves in (b) represent two distributions with the same dispersion but with unequal means. $\overline{X}_1$ and $\overline{X}_2$. (c) represents two distributions with unequal dispersion.

The measures of central tendency are, therefore, insufficient. They must be supported and supplemented with other measures. In this chapter, we shall be especially concerned with the measures of variability, or spread or dispersion. A measure of variation or dispersion is one that measures the extent to which there are differences between in dividable observation and some central or average value. In measuring variation we shall be interested in the amount of the variation or its *decrepit* not in the *direction*. For example, a measure of 6 inches below the mean has just as much dispersion as a measure of six inches above the mean.

Significance of Measuring Variation

Measures of variation are needed for four basic purposes:

1. To Compare Two or More Series with Regard to their Variability
2. To Facilitate the use of Other Statistical Measures.
3. To determine the Reliability of an Average.
4. To serve as a basis for the control of the variability.

A brief explanation of these points is given as follows:

(i) Measures of dispersion enable a comparison to be made of two or more series with regard to their variability. The study of variation may also be looked upon as a means of determining uniformity of consistency. A high degree of variation would mean little uniformity or consistency whereas a low degree of variation would mean great uniformity or consistency.

(ii) Many powerful analytical tools in statistics such as correlation analysis, the testing of hypothesis, analysis of variance, the statistical quality control, regression analysis are based on measures of variation of one kind or another.

(iii) Measures of variation point out as to how far an average is representative of the mass. When dispersion is small. The average is a typical value in the sense that it closely represents the individual value and it is reliable in the sense that it is a good estimate of the average in the corresponding universe. On the other hand, when dispersion is large, the average is not so typical, and unless the sample is very large, the average may be quite unreliable.

(iv) Another purpose of measuring dispersion is to determine nature and cause of variation in order to control the variation itself. In matters of health variations in body temperature, pulse beat and blood pressure are the basic guides to diagnosis. Prescribed treatment is designed to control their variation. In industrial production efficient operation requires control of quality variation, the causes of which are sought through inspection is basic to the control of causes of specially important. In social sciences a special problem requiring the measurement of variability is the measurement of "inquality" of the distribution of income or wealth, etc.

Properties of A Good Measure of Variation

A good measure of dispersion should possess, as far as possible, the following properties:

(i) It should be based on each and every item of the distribution.

(ii) It should be amenable to further algebraic treatment.

(iii) It should be have sampling stability.

(iv) It should be simple to understand.

(v) It should be easy to compute.

(vi) It should be rigidly defined.

(vii) It should not be unduly affected by extreme items.

Methods of Studying Variation

The following are the important methods of studying variation:

1. The Range.
2. The Interquartile Range and the Quartile Deviation.
3. The Mean Deviation or Average Deviation.
4. The Standard Deviation, and
5. The Lorenz Curve.

Of these the first two, namely, the range and quartile deviations, are positional measures because they depend on the values at a particular position in the distribution. The other two, the average deviation and the standard deviation, are called calculation measures of deviation because all of the values are employed in their calculation and the last one is a graphic method.

Absolute and Relative Measures of Variation

Absolute measures of dispersion are expressed in the same statistical unit in which the original data are given such as rupees, kilograms, tonnes etc. These values may be used to compare the variations in two distributions provided the variables are expressed in the same units and of the same average size. In case the two sets of data are expressed in different units, however, such as quintals of sugar versus tonnes of sugarcane, or if the average size is very different such as manager's salary versus workers' salary, the absolute measures of dispersion are not comparable. In such cases measures of relative dispersion should be used.

A measure of relative dispersion is the ration of a measure of absolute dispersion to an appropriate average. It is sometimes called a coefficient of dispersion, because "coefficient" means a pure number that is in dependent of relative dispersion the average used as base should be the same one from which the absolute deviations were measured.

Range is the simplest method of studying dispersion. It is defined as the difference between the value of the smallest item and the value of the largest item included in the distribution. Symbolically, we have

$$\text{Range} = L - S$$

where L = Largest item, and

S = Smallest item.

The relative measure corresponding to range, called the coefficient of range, is obtained by applying the following formula:

$$\text{Coefficient of Range} = \frac{L - S}{L + S}.$$

Example 1:

Compute coefficient of quartile deviation from the following data:

Marks	*10*	*20*	*30*	*40*	*50*	*80*
No. of Students	*4*	*7*	*15*	*8*	*7*	*2*

Solution:

The Calculation of coefficient of quartile deviation is calculated as follows:

Marks	***Frequency***	***c.f.***	***Marks***	***Frequency***	***c.f.***
10	4	4	40	8	34
20	7	11	50	7	41
30	15	26	60	2	43

Here we have

$$Q_1 = \text{Size of } \frac{N+1}{4} \text{ th item} = \frac{43+1}{4} = 11\text{th item.}$$

Size of 11th item is 20. Thus $Q_1 = 20$

$$Q_3 = \text{Size of } 3\left(\frac{N+1}{4}\right) \text{ th item} = \frac{3 \times 44}{4} = 33\text{rd item.}$$

Size of 33rd item is 40. Thus $Q_3 = 40$

$$\text{Q.D.} = \frac{Q_3 - Q_1}{2} = \frac{40 - 20}{2} = 10$$

$$\text{Coefficient of Q.D.} = \frac{Q_3 - Q_1}{Q_3 + Q_1} = \frac{40 - 20}{40 + 20} = 0.333$$

$$Q_3 = \text{Size of } \frac{3N}{4} \text{th item } \frac{3 \times 200}{4} = 150\text{th item}$$

Q_3 lies in the class 38 – 40

we have $$Q_3 = L + \frac{3N/4 - \text{c.t.}}{f} \times i$$

$$L = 38,\ 3N/4 = 150,\ c.f. = 76,\ t = 99,\ i = 2$$

$$Q_3 = 38 + \frac{150 - 76}{99} \times 2 = 38 + 1.49 = 39.\ 49$$

$$Q.D. = \frac{39.49 - 36.16}{2} = 1.67$$

$$\text{Coefficient of Q.D} = \frac{Q_3 - Q_1}{Q_3 + Q_1} = \frac{39.49 - 36.16}{39.49 + 36.16} = \frac{3.33}{75.65} = 0.044$$

MERITS AND LIMITATIONS

Merits

(a) It is also useful in erratic or badly skewed distributions where the other measures of dispersion would be warped by extreme values. The quartile deviation is not affected by the presence of *extreme values.*

(b) It has a special utility in measuring in case of open end distributions or one in which the data may be ranked but measured quantitatively.

Limitations

As the value of quartile deviation does not depend upon every item of the series it cannot be regarded as a good method of measuring dispersion.

(a) It is in fact not a measure of dispersion as it really does not show the scatter around an average but rather a distance on a scale, *i.e.*, quartile deviation is not itself measured from an average, but it is a positional average. Consequently, some statisticians speak of quartile deviation as a measure of *partition* rather than a measure of dispersion. If we really desire to measure variation in the sense of showing the scatter round an average, we must include the deviation of each and every item from an average in the measurement.

(b) It is not capable of mathematical manipulation.

(c) Its value is very much affected by sampling fluctuations.

Percentile Range

Like semi-interquartile range, the percentile range is also used as a measure of dispersion. Percentile range of a set of data is defined as:

$$\text{Percentile Range} = P_{90} - P_{10}$$

where P_{90} and P_{10} are the 90th and 10th percentiles respectively. The semi-percentile range, *i.e.* $\left(\frac{P_{90} - P_{10}}{2}\right)$ can also be used, but is not commonly employed

THE MEAN ELEVATION

To study the formation of a distribution we should take the deviations from an average. The two other measures namely, the average deviation and the standard deviation help us in achieving this goal.

The mean deviation is also known as the average deviation. It is the average difference between the items in a distribution and the median or mean of that series. Theoretically there is an advantage in taking the deviations from median because the *sum of deviations of items from median is minimum when signs are ignored.* However, in practice, the arithmetic mean is more frequently used in calculating the value of average deviation and this is the reason why it is more commonly called mean deviation.

Computation of Mean Deviation-Individual Observations

If X_2, X_1, X_3, X_N are N given observations then the deviation about an average A is given by

$$\text{M.D.} = \frac{1}{n} \Sigma \mid X - A \mid \frac{1}{N} \Sigma \mid D \mid \quad \Rightarrow \quad \frac{\Sigma \mid D \mid}{N}$$

where $\mid D \mid = X - A \mid$. Read as mod (X – A) is the modulus value or absolute value of the deviation ignoring plus and minus signs.

Steps

(a) Obtain the total of these deviations, *i.e.*, $\Sigma \mid D \mid$.

(b) Compute the median of the series.

(c) The deviations of items from median ignoring ± signs and denote these deviations by $\mid D \mid$.

(d) Divide the total obtained in step (iii) by the number of observations.

If a distribution is normal, the mean ± mean deviation is the range that will include 57.7 percent of the items in the series. If it is moderately skewed, then we may expect approximately 57.5 percent of the items to fall within this range. Hence, if average deviation is small, the distribution is highly compact or uniform, since more than half of the cases are concentrated within a small range around the mean.

Example 2:

The mean and the standard deviation of one sample are respectively 54.4 and 8; the mean and the standard deviation of another sample are 50.3 and 7 respectively. The size of the first sample is 50 and that of the second is 1000. Find the mean and standard deviation of the composite sample (size 150) combining the aforesaid to samples.

Solution:

We are given the following information:

$\overline{X}_1 = 54.4, \sigma_1 = 8, n_1 = 50 \ \overline{X}_2 = 50.3, \sigma_2 = 7, n_2 = 100$

We have to find $\overline{X}_{12}$ and σ_{12} $\overline{X}_{12} = \dfrac{n_1 \overline{X}_1 + n_2 \overline{X}_2}{n_1 + n_2}$

$$= \frac{(50 \times 54.4) + (100 \times 50.3)}{50 + 100} = \frac{2720 + 5030}{150} = \frac{7750}{150} = 51.67$$

$$\sigma_{12} = \sqrt{\frac{n_1\sigma_1^2 + n_2\sigma_2^2 + n_1\sigma_1^2 + n_2\sigma_2^2}{n_1 + n_2}}$$

$$d_1 = |\overline{X}_1 - \overline{X}_{12}| = |54.4 - 51.67| = 2.73$$

$$d_2 = |\overline{X}_2 - \overline{X}_{12}| = |50.3 - 51.67| = 1.37$$

$$\sigma_{12} = \sqrt{\frac{50(8)^2 + 100(7)^2 + 50(2.73)^2 + 100(1.37)^2}{150}}$$

$$= \sqrt{\frac{3200 + 4900 + 372.645 + 187.69}{150}} = \sqrt{\frac{8660.335}{150}} = 7.6$$

Example 3:

Two brands of tyres are tested with the following results:

Life (in '000 miles)	*No. of tyres brand*	
	X	*Y*
20-25	*1*	*0*
25-30	*22*	*24*
30-35	*64*	*76*
35-40	*10*	*0*
40-45	*3*	*0*

(a) *Which brand of tyres have greater average life?*

(b) *Compare the variability and state which brand of tyres would you use on your fleet of trucks?*

Solution:

In order to answer part :

(a) we have to compare the means and to answer part

(b) compare the coefficient of variation.

Calculation of Coefficient of Variation (brand X)

Life *('000 miles)*	*m.p.* *m*	*f*	*(m-32.5)/5* *d*	*fd*	*fd*^{2}
20-25	22.5	1	–2	–2	4
25-30	27.5	22	–1	–22	22
30-35	32.5	64	0	0	0
35-40	37.5	10	+1	+10	10
40-45	42.5	3	+2	+6	12
		N = 100		$\sum$fd = – 8	$\sum$fd^2 = 48

$$\overline{X} = A + \frac{\sum fd}{N} \times i$$

A = 32.5, $\sum$fd = –8,

N = 100, i = 5

$$\overline{X} = 32.5 - \frac{8}{100} \times 5 = 32.5 - 4 = 32.1$$

$$\sigma = \sqrt{\frac{\sum fd^2}{N} - \left(\frac{\sum fd}{N}\right)^2} \times i = \sqrt{\frac{48}{100} - \left(\frac{-8}{100}\right)^2} \times 5$$

$$= \sqrt{4. - 0064} \times 5 = 6274 \times 5 = 3.137$$

$$C.V. = \frac{\sigma}{X} \times 100 = \frac{3.137}{32.1} \times 100 = 9.773$$

Calculation of Coefficient of Variation (brand X)

Life *('000 miles)*	*m.p.* *m*	*f*	*(m-32.5)/5* *d*	*fd*	*fd*^{2}
20-25	22.5	0	–2	–2	0
25-30	27.5	24	–1	–24	24
30-35	32.5	76	0	0	0
35-40	37.5	0	+1	0	0
40-45	42.5	0	+2	0	0
		N = 100		$\sum$fd = – 24	$\sum$fd^2 = 48

$$\overline{X} = A + \frac{\sum fd}{N} \times i$$

$A = 32.5$, $\sum fd = -24$, $N = 100$, $i = 5$

$$\overline{X} = 32.1 - \frac{24}{100} \times 5 = 32.5 - 12 = 31.\ 3$$

$$\sigma = \sqrt{\frac{\sum fd^2}{N} - \left(\frac{\sum fd}{N}\right)^2} \times i \sqrt{\frac{24}{100} - \left(\frac{-24}{100}\right)^2} \times 5$$

$$= \sqrt{24 - 0576} \times 5 = 1824 \times 5 = 3.137$$

$$\text{C.V.} = \frac{\sigma}{X} \times 100 = \frac{0.912}{31.3} \times 100 = 2.914$$

(a) Since arithmetic mean is more for brand X of tyres, they have greater average life.

(b) Since coefficient of variation is less for brand Y of tyres, they are more consistent and should be preferred for use.

Example 4:

The mean and standard deviation of a set of 100 observations were worked out as 40 and 5 respectively by a computer which by a computer which by mistake took the value 50 in place of 40 one observation. Find the correct mean and variance.

Solution:

$$\overline{X} = \frac{\sum X}{N},\ N\overline{X}\ \sum X,\ N = 100,\ \overline{X} = 40$$

$$\sum X = 100 \times 40 = 4{,}000$$

But this is not the correct $\sum X$ because one item has been taken as 50 instead of 40.

$\therefore$ Correct $\sum X = 4000 - 50 + 40 = 3{,}990$

$$\text{Correct Mean} = \frac{3990}{100} = 39.90$$

Correct variance

$$\text{Variance} = \frac{\sum X^2}{N} - (\overline{X})^2$$

$$\text{Variance} = \sigma^2 = (5)^2 = 25,\ N = 100$$

$$25 = \frac{\sum X^2}{100} - (40)^2$$

$$2500 = \Sigma X^2 - 160000 \quad \Rightarrow \quad \Sigma X^2 = 160000 + 2500 = 162500$$

Correct $\Sigma X^2 = 162500 - (50)^2 + (40)^2 = 162500 - 2500 + 1600 = 161600$

$$\text{Correct Variance} = \frac{\text{Correct} \sum X^2}{N} - (\text{Correct } \overline{X})^2$$

$$= \frac{161600}{100} - (39.9)^2 = \frac{161600 - 159201}{100} = \frac{2399}{100} = 23.99$$

Thus, correct mean = 39.9 and correct Variance = 23.99

Example 5:

The mean of 5 observations is 4.4 and the variance is 8.24. If the three of the five observations are 1, 2 and 6, find the other two.

Solution:

$$\overline{X} = \frac{\sum X}{N}, \quad \Sigma X = N\overline{X}$$

Here $N = 5$, $\overline{X} = 4.4$; $\Sigma X = 5 \times 4.4 = 22$

Let the two missing items be x_1 and x_2

$\therefore 1 + 2 + 6 + x_1 + x_2 = 22$

$\Rightarrow \quad x_1 + x_2 = 22 - 9$

$\Rightarrow \quad x_1 + x_2 = 13$

$$\sigma^2 = \frac{\sum X^2}{N} - \overline{X}^2$$

$$\Rightarrow \quad 8.24 = \frac{\sum X^2}{5} - (4.4)^2$$

$41.2 = \Sigma X^2 - 19.36 \times 5$

$\Rightarrow \quad \Sigma X^2 = 96.80 + 41.20 = 138$

$\Sigma X^2 = x_1 + x_2^2 + 1^2 + 2^2 + 6^2 = x_1^2 + x_2^2 + 41$

$x_1^2 + x_2^2 = 138 - 41 = 97$

$(x_1 + x_2)^2 = x_1^2 + x_2^2 + 2x_1x_2$

$(13)^2 = 97 + 2x_1x_2$

$\Rightarrow \quad x_1x_2 = 36$

$2x_1x_2 = 169 - 97$

$\Rightarrow \quad x_1x_2 = 36$

$(x_1 - x_2)^2 = x_1^2 + x_2^2 - 2x_1x_2 = 97 - 2(36) = 25$ ...(i)

$x_1 - x_2 = 5$

Adding $\underline{x_1 + x_2} = 13$ ...(ii)

$2x_1 = 18$

$x_2 = 9$ ...(iii)

Putting the value of x_1 in equation (i)

$$9 + x_2 = 13 \quad \therefore \quad x_2 = 4$$

Thus the two missing values are $x_1 = 9$, $x_2 = 4$.

Example 6:

The following are the prices of shares of AB Co. Ltd. from Monday to Saturday

Day	***Price (Rs.)***	***Day***	***Price (Rs.)***
Monday	200	Thursday	160
Tuesday	210	Friday	220
Wednesday	208	Saturday	250

Calculate range and its coefficient.

Solution:

Range = L – S

Here L = 250 and S = 160

Range = 250 – 160 = Rs. 90

$$\text{Coefficient of Range} = \frac{L - S}{L + S} = \frac{250 - 160}{250 + 160} = \frac{90}{160} = 0.22.$$

Continuous Series

There are two methods of determining the range from data grouped into a frequency distribution. The first method is to find the difference between the upper limit of the highest wage class and the lower limit of the lowest wage class. The other method is to subtract the midpoint of the lowest wage class from the mid-point of the highest wage class. In practice, both the methods are used.

Example 7:

Calculate the mean deviation and its coefficient of the two income groups of five and seven members given below:

I (Rs.): 4,000 4,200 4,400 4,600 4,800

II (Rs.): 3,000 4,000 4,200 4,400 4,600 4,800 5,800

Solution:

Calculation of Mean Deviation

	Deviation form median 4400 | D |		*Deviation from median 4400 | D |*
4,000	400	3,000	1,400
4,200	200	4,000	400
4,400	0	4,200	200
4,600	200	4,600	0
4,800	400	4,600	200
		4,800	400
		5,800	1,400
n = 5	$\Sigma \mid D \mid = 1200$	N = 7	$\Sigma \mid D \mid = 4,000$

Mean Deviation: Group I: $\text{M.D.} = \dfrac{\sum \mid D \mid}{N}$

$\mid D \mid$ = Deviation from median ignoring signs,

$$\text{Median} = \text{Size of } \frac{N+1}{2}\text{th item} = \frac{5+1}{2} = \text{3rd item}$$

Size of 3rd item is 4,400 $\quad \text{M.D.} = \dfrac{1,200}{5} = 240$

This means that the average deviation of the individual incomes from the median income is Rs. 240.

Mean Deviation: Group II

$$\text{Mean} = \text{Size of } \frac{N+1}{2}\text{th item} = \frac{7+1}{2} = \text{4th item}$$

Size of 4th item is 4, 400

$$\Sigma \mid D \mid = 4,000, \quad N = 7.$$

$$\text{M.D.} = \frac{4,000}{7} = 571.43.$$

Note: If we were to compute coefficient of mean deviation we shall divide mean deviation by median. Thus for the first group:

$$\text{Coefficient of M.D.} = \frac{240}{4,400} = 0.\ 054$$

and for the second group

$$\text{Coefficient of M.D.} = \frac{571.43}{4,400} = 0.\ 130.$$

Calculation of Mean Deviation

Discrete Series : In discrete series the formula for calculating mean deviation is

$$\text{M.D.} = \frac{\sum f|D|}{N} \text{ (by the same logic as given before)}$$

| D | denotes deviation from median ignoring signs.

Steps.

(a) Multiply these deviations by the respective frequencies and obtain the total $\sum f|D|$.

(b) Take the deviations of the items from median ignoring signs and denote them by | D |.

(c) Divide the total obtained in Step (ii) by the number of observations. This gives us the value of mean deviation.

(d) Calculate the median of the series.

Example 8:

Find out the value of quartile deviation and its coefficient from the following data:

Roll No.	*1*	*2*	*3*	*4*	*5*	*6*	*7*
Marks	*20*	*28*	*40*	*12*	*30*	*15*	*50*

Solution:

Calculation of Quartile Deviation

Marks arranged in ascending order: 12 15 20 28 30 40 50

$$Q_1 = \text{Size of } \frac{N+1}{4}\text{th item} = \text{Size of } \frac{7+1}{4} = \text{2nd item}$$

Size of 2nd item is 15. Thus $Q_1 = 15$

$$Q_3 = \text{Size of } 3\left(\frac{N+1}{4}\right)\text{th item} = \text{Size of } \left(\frac{3 \times 8}{4}\right)\text{th item} = \text{6th item}$$

Size of 6th item is 40. Thus $Q_3 - 40$

$$\text{Q.D.} = \frac{Q_3 - Q_1}{2} = \frac{40 - 15}{2} = 12.\ 5.$$

$$\text{Coefficient of Q.D.} = \frac{Q_3 - Q_1}{Q_3 + Q_1} = \frac{40 - 15}{40 + 15} = \frac{25}{55} = 0.455.$$

Example 9:

Calculate quartile deviation and the coefficient of quartile deviation from the following data:

Wages in Rupees per week	*less then 35*	*35-37*	*38-40*	*41-43*	*over 43*
Number of wage earners	*14*	*62*	*99*	*18*	*7*

Solution:

Calculation of Q.D. and its coefficient

Wages (Rs. per week)	***f***	***c.f.***
Less than 35	14	14
35–37	62	76
38–40	99	175
41–43	18	183
over 43	7	200

$$\text{Q.D.} = \frac{Q_3 - Q_1}{2}$$

$$Q_1 = \text{Size of } \frac{N}{4}\text{th item} = \frac{200}{4} = 50 \text{ th item}$$

Q_1 lies in the class 35–37.

$$Q_1 = L + \frac{N/4 - \text{c.t.}}{f} \times i$$

$$L = 35,\ N/4 = 50,\ \text{c.f.} = 14,\ f = 62,\ i = 2$$

$$Q_1 = 35 + \frac{50 - 14}{62} \times 2 = 35 + 1.16 = 36.16$$

The relative measure corresponding to the mean deviation, called the coefficient of mean deviation, is obtained by dividing mean deviation by the particular average used in computing mean deviation. Thus if mean deviation has been computed from median, the coefficient of mean deviation shall be obtained by dividing mean deviation by median.

$$\text{Coefficient of M.D.} = \frac{\text{M.D}}{\text{Median}}.$$

If mean has been used while calculating the value of mean deviation, in such a case coefficient of mean deviation shall be obtained by dividing mean deviation by the mean.

Example 9(a):

Calculate Coefficient of Range from the following data:

Marks	*No. of students*	*Marks*	*No. of students*
10–20	*8*	*40–50*	*8*
20–30	*10*	*50–60*	*4*
30–40	*12*		

Solution:

$$\text{Coefficient of Range} = \frac{L - S}{L + S} = \frac{60 - 10}{60 + 10} = \frac{50}{70} = 0.714$$

Merits and Limitations: The merits and limitations of Range can be numerated here.

Imitations.

(a) Range cannot tell us anything about the character of the distribution within the two extreme observations. Observe the following three series:

Series A	46	6	46	46	46	46	46	46
Series B	6	10	6	6	46	46	46	46
Series C	6	6	15	25	30	32	40	46

In all the three series range is the same, *i.e.*, (46–6) = 40. But it does not mean that the distributions are alike. The range takes no account of the form of the distribution within the range. Range is, therefore, unreliable as a guide to the dispersion of the value within a distribution.

(b) It is subject to fluctuations of considerable magnitude from sample to sample.

(c) Range is not based on each and every item of the distribution.

Merits.

(a) It takes minimum time to calculate the value of range. Hence, if one is interested in getting a quick rather than very accurate picture of variability one may compute range.

(b) Amongst all the methods of studying dispersion range is the simplest to understand and the easiest to compute.

Uses : Despite serious limitations range is useful in the following cases:

(i) *Quality Control.* The object of quality control is to keep a check on the quality of the product without 100% inspection. When statistical methods of quality control are used, control charts are prepared and in preparing these charts range plays a very important role. The idea basically is that if the range-the difference between the largest and smallest mass produced items-increases beyond a certain point. The production machinery should be examined to find out why the items produced have not followed their usual more consistent pattern.

(ii) *Fluctuations in the Share Prices.* Range is useful in studying the variations in the prices of stocks and shares and other commodities that are sensitive to price changes from one period to another. For example, by computing range we can get an idea about the range of variation of say, gold prices. If the minimum price for 10 gm. gold in the year 1999 was Rs. 4050 and the maximum price Rs. 4950 this at once tells us about the range of variation, *i.e.*, Rs. 900 (4950 – 4050).

(iii) *Weather Forecasts.* The meteorological department does make use of the range in determining, say, the difference between the minimum temperature and the maximum temperature. This information is of great concern to the general public because they know as to within what limits the temperature is likely to vary on a particular day.

(iv) *Everyday Life.* The range is a most commonly used measure of dispersion in everyday life. Questions of the form "what is the minimum and maximum temperature on a particular day"? "What is the difference between the wages earned by workers of a particular factory"? How much one spends on petrol in his car/scooter in a month"–are all usually answered in the form of range. Answers to questions such as these are usually given in the form of 'Between such and such. Regardless of the crudity of expression the answer is still a range.

The Interquartile Range or The Quartile Deviation

If the dispersion of the extreme items is discarded, the limited range thus established might be more instructive. For this purpose there has bee developed a measure called the *interquartile range,* the range which includes the middle 50 percent of the distribution. That is one quarter of the observations at the lower end, another quarter of the observations at the upper end of the distribution are excluded in computing the interquartile range. In other words, interquartile range represents the difference between the third quartile and the first quartile.

$$\text{Interquartile range} = Q_3 - Q_1.$$

Very often the interquartile range is reduced to the form of the Semi-interquartile range or quartile deviation by dividing it by 2.

$$\text{Quartile Deviation or Q.D} = \frac{Q_3 - Q_1}{2}.$$

Quartile deviation gives the average amount by which the two quartiles differ from the median. In symmetrical distribution the two quartiles (Q_1 and

Q_3) are equidistant from the median, *i.e.*, Med. – Q_1 = Q_3 – Med. and as such the difference can be taken as a measure of dispersion. The median ± Q.D. covers exactly 50 percent of the observations.

In reality, however, one seldom finds a series in business and economic data that is perfectly symmetrial. Nearly all distributions of social series are asymmetrical. In an asymmetrical distribution. Q_1 and Q_3 are not equidistant from the median. As a result an asymmetrical distribution includes only approximately 50 percent of observations.

When quartile deviation is very small, it describes high uniformity or small variation of the central 50% items and a high quartile deviation means that the variation of the central 50% items and a high quartile deviation means the variation among the central items is large.

Quartile deviation is an absolute measure of dispersion. The relative measure corresponding to this measure, called the coefficient of quartile deviation, is calculated as follows.

$$\text{Coefficient of Q.D.} = \frac{(Q_3 - Q_1)/2}{(Q_3 + Q_1)/2} = \frac{Q_3 - Q_1}{Q_3 + Q_1}$$

Coefficient of quartile deviation can be used to compare the degree of variation in different distributions.

Computation of Quartile Deviation

The process of computing quartile deviation is very simple, we have just to compute the values of the upper and lower quartiles. The following illustrations would clarify calculations.

Example 10:

X Ltd. is actively considering the following two mutually exclusive project for adoption.

Year	*Project X* *Cost profit (Rs. in Lakhs)*	*Project Y* *Cash profit (Rs. in Lakhs)*
1	*10*	*5*
2	*5*	*25*
3	*20*	*45*
4	*40*	*30*
5	*60*	*30*

Which is the most risky project (use coefficient of variation)

Solution:

For finding out which is more risky project out of X and Y, we compare coefficent of variation.

Calculation of Coefficient of Variation

Project X			*Project Y*		
X	$(X - \overline{X})$ *x*	x^2	*Y*	$(Y - \overline{Y})$ *y*	y^2
10	–17	289	5	–22	484
5	–22	484	25	–2	4
20	–7	49	45	+18	324
40	+13	169	30	+3	9
60	+33	1089	30	+3	9
$\Sigma x = 135$	$\Sigma x = 0$	$\Sigma x^2 = 2080$	$\Sigma y = 135$	$\Sigma y = 0$	$\Sigma y^2 = 830$

Project X

$$C.V. = \frac{\sigma}{X} \times 100$$

$$\overline{X} = \frac{\sum x}{N} = \frac{135}{7} = 27$$

$$\sigma = \sqrt{\frac{\sum x^2}{N}} = \sqrt{\frac{2080}{5}} = 20.4$$

$$C.V. = \frac{20.4}{27} \times 100 = 75.56$$

Project Y

$$C.V. = \frac{\sigma}{Y} \times 100$$

$$\overline{Y} = \frac{\sum y}{N} = \frac{135}{7} = 27$$

$$\sigma = \sqrt{\frac{\sum y^2}{N}} = \sqrt{\frac{830}{5}} = 12.88$$

$$C.V. = \frac{12.88}{27} \times 100 = 47.7$$

Since coefficient of variation is much more for project X hence it is a more risky project.

Which Measure of Dispersion to Use

The choice of a suitable measure depends on the following two factors:

(i) *The type of data available.* If they are few in number, or contain extreme values, avoid the standard deviation. If they are generally skewed, avoid the mean deviation as well. If they have gaps around the quartiles, the quartile deviation should be avoided. If there are open-end classes, the quartile measure of dispersion should be preferred.

(ii) *The purpose of investigation.* In an elementary treatment of statistical series in which a measure of variability is desired only for itself any of three measures, namely, range, quartile deviation and average deviation, would be acceptable. Probably the average deviation would be better. However, in usual practice, the measure of variability is employed in further statistical analysis. For such a purpose the standard deviation, by far, is the most popularly used. It is free from those defects from which other measures suffer. It leads itself to the analysis of variability in terms of normal curve of error. Practically all advanced statistical methods deal with variability and centre around the standard deviation. Hence, unless the circumstances warrant the use of any other measure, we should make use of standard deviation for measuring variability.

Example 11:

You are given below the daily wages paid to the workers in two factories X and Y:

Daily Wages	***No. of workers***	
	Factory X	***Factory Y***
12-13	*15*	*25*
13-14	*30*	*40*
14-15	*44*	*60*
15-16	*60*	*35*
16-17	*30*	*12*
17-18	*14*	*15*
18-19	*7*	*5*

Using appropriate measure answer the following?

(i) *Which factory pays higher average wage?*

(ii) *Which factory has more consistent wage structure.*

Solution:

For finding out which factory pays higher average wage, we have to compute the arithmetic means and for finding out which factory has more consistent wage structure, we have to compare coefficient of variation.

Calculation of $\overline{X}$ and C.V.

Wages (Rs.)	*m.p*	*t*	Factory X *(m – 15)/3*	*fd*	*fd²*	Factory Y *f*	*fd*	*fd²*
12-13	12.5	15	–3	–45	135	25	–75	225
13-14	13.5	30	–2	–60	120	40	–80	160
14-15	14.5	44	–1	–44	44	60	–60	60
15-16	15.5	60	0	0	0	35	0	0
16-17	16.5	30	+1	+30	30	12	+12	12
17-18	17.5	14	+2	+28	56	15	+30	60
18-19	18.5	7	+3	+21	63	5	+15	45
		N = 200		Σfd = – 70	Σfd² = 448		Σfd = –- 158	Σfd² = 562

Factory X factory Y

$$\overline{X} = A + \frac{\sum fd}{N} = 15.5 - \frac{70}{200} = 15.15$$

$$\overline{X} = A + \frac{\sum fd}{N} = 15.5 - \frac{158}{200} = 14.71$$

Since arithmetic mean is higher for factory X, hence factory X pays higher average wage.

$$\sigma = \sqrt{\frac{\sum fd^2}{N} - \left(\frac{\sum d}{N}\right)^2} \qquad \sigma = \sqrt{\frac{\sum fd^2}{N} - \left(\frac{\sum fd}{N}\right)^2}$$

$$= \sqrt{\frac{448}{200} - \left(\frac{-70}{200}\right)} \qquad = \sqrt{\frac{562}{200} - \left(\frac{-158}{200}\right)^2}$$

$$= \sqrt{2.24 - 1225} = 1.455 \qquad = \sqrt{2.81 - 624} = 1.479$$

$$C.V. = \frac{\sigma}{X} \times 100 = \frac{1.45}{15.15} = 9.60$$

$$C.V. = \frac{\sigma}{X} \times 100 = \frac{0.479}{14.71} \times 100 = 10.5$$

Since coefficent of variation is less for factory X, hence factory X has more consistent wage.

TCHEBYCHEFF'S THEOREM

According to this theorem, given a group of N numbers, at least the proportion $1 - (1/K)^2$ of the N observations will lie within K standard deviations of the mean. The symbol K represents number of standard deviations. The theorem can be quantified for any desired value of K. Each different value of K produces a new interval with different minimum proportions of observations encompassed.

To illustrate let K = 1.2 and 3. When K = 1, this interval created is the mean ± 1 standard deviation, *i.e.*, $\mu \pm \sigma$. For this value of K. The minimum proportion of observations contained in the interval is

$$1 + (1/K)^2 = 1 - 1/1 = 1 - 1 = 0$$

When K = 2. Then interval is $\mu \pm 2\sigma$ and the minimum proportion of items in the interval is:

$$1 - (1/3)^2 = 1 - 1/9 = 8/9$$

As K increases, the fraction obtained above would continue to increase as can be seen from the following table:

Tchebycheff Proportions

K values	*Ranges*	*Minimum Proportion of Observations*			
1	$\mu \pm \sigma$			0	
2	$\mu \pm 2\sigma$	3/4	or	75	percent
3	$\mu \pm 3\sigma$	8/9	or	88.89	percent
4	$\mu \pm 4\sigma$	15/16	or	93.75	percent
5	$\mu \pm 5\sigma$	24/25	or	96.00	percent
:	:		:		
K	$\mu \pm k\sigma$		$1 - (1/K)^2$		

Lorenz Curve

The lorenz Curve, devised by Max O. Lorenz, famous economic statistician, is a graphic method of studying dispersion. This curve was used by him for the first time to measure the distribution of wealth and in wages, turnover, etc. However, still the most common use of this curve is in the study of the degree of inequality in the distribution of income and wealth between countries or between different periods of time. It is a cumulative percentage curve in which the percentage of items is combined with the percentage of other things as wealth. profits, turnover, etc.

While drawing the Lorenz curve the following procedure is adopted:

(i) The size of items (variable values) and frequencies are both cumulated. Taking grand total for eachas 100, percentages are obtained for these various coumulative values.

(ii) On the X-axis start from 0 to 1000 and take the percent of comulative frequencies.

(iii) On the Y-axis start from 0 to 100 and take the percent of the cumulated values of the variable.

(iv) Draw a disgonal line joing O(0, 0) with the point P(100, 100) as shown in the diagram below. The line OP will make an angle of 45° with the Y-axis and is called the line of equal distribution. Any point on this diagonal shows that same percent on X as on Y.

(v) Plot the percentages of the cumulated values of the variable (Y) against percentages of the correseponding cumulated frequencies (X) for the given distribution and join these points with a smooth freehand curve. For any given distribution this will never cross the line of equal distribution OP. It will always lie below OP unless the distribution is uniform in which case it will coincide with OP. The greater the variability, the greater is the distance of the curve from OP.

In the above diagram OP is the line of equal distribution. The points lying on the curve OAP indicate a less degree of variability as compared to the points lying on the curve OBP. When the points lie on the curve OCP. variability is still greater. Thus a measure of variability of the distribution is provided by the distance of the curve of the cumulated percentages of the given distribution from the line of equal distribution.

Example 12:

A student obtained the mean and standard deviation of 100 observations as 40 and 5.1 respectively. It was later found that one observation was wrongly copied as 50, the correct figure being 40. Find the correct men and standard deviation.

Solution:

We are given $\overline{X} = 40, \sigma = 5.1, N = 100$

$$\overline{X} = \frac{\sum X}{N}$$

$$40 = \frac{\sum X}{100} \Rightarrow \Sigma X = 4{,}000$$

But correct $\Sigma X = \Sigma X$ – wrong items + correct items

$= 4{,}000 - 50 + 40 = 3990$

$\therefore$ Correct $= \overline{X} = \frac{\text{Correct}\sum X}{N} = \frac{3990}{100} = 39.9$

Corect standard deviation

$$\sigma = \sqrt{\frac{\sum X}{N} - (\overline{X})^2} \Rightarrow 5.1 = \sqrt{\frac{\sum X^2}{100} - (40)^2}$$

Squaring $26.01 = \frac{\sum X^2}{100} - 1600$

$\Rightarrow$ $2601 = \sum X^2 - 1{,}60{,}000$ $\sum X^2 = 162601$

Correct $\sum X^2 = 162601 - (50)^2 + (40)^2 = 162601 - 2500 + 1600 = 161701$

$$\text{Correct } \sigma = \sqrt{\frac{\text{Correct} \sum X}{N} - (\text{Correct } \overline{X})^2} = \sqrt{\frac{161701}{100} - (39.9)^2}$$

$$= \sqrt{161701 - 1592.02} = \sqrt{25} = 5.$$

Example 13:

Find the interquartile range and the coefficient of quartile deviation from the following data:

Markes in Staistics:above	*0*	*10*	*20*	*30*	*40*	*50*	*60*	*70*	*80*
No. of Students:	*150*	*140*	*100*	*80*	*80*	*70*	*30*	*14*	*0*

Solution:

Calculation of Interoquartile Range and Coefficient of Q.D.

Maks	*f*	*c.f*
0.10	10	10
10-20	40	50
20-30	20	70
30-40	0	70
40-50	10	80
50-60	40	120
60-70	16	136
70-80	14	150

Interquartile Range $= Q_3 - Q_1$

Q_1 Size of $\frac{N}{4}$th item $= \frac{150}{4} = 37.5$th item

Q_1 lies in the class 10-20

$$Q_1 = L + \frac{N/4 - c.f}{f} \times i$$

L = 10, N/4 = 375, c.f 10, f = 40, i = 10

$$Q_1 = 10 + \frac{37.5 - 10}{40} \times 10 = 10 + 6.875 = 16.875$$

$$Q_3 = \text{Size of } \frac{3N}{4}\text{th item} = \frac{3 \times 150}{4} = 112.5\text{th item}$$

Q_3 lies in the class 50-60

$$Q_3 = L + \frac{3N/4 = c.f.}{f} \times i$$

L = 50, 3N/4 = 112.5, c.f. = 80, f = 40, i = 10

$$Q_3 = 50 + \frac{112.5 - 80}{40} \times 10 = 50 + 8.125 = 58.125$$

$$Q_3 - Q_1 = 58.125 - 16.875 = 45.623$$

$$\text{Coeff. of Q.D.} = \frac{Q_3 - Q_1}{Q_3 + Q_1} = \frac{58.125 - 16.875}{58.125 + 16.875} = \frac{45.625}{75} = 0.608$$

Example 14:

From the following information, find the standard deviation of x and y variable:

$\Sigma x = 235, \quad \Sigma x = 250$

$\Sigma x^2 = 6750, \quad \Sigma y^2 = 6840$

$N = 10$

Solution:

$$\text{S.D. of } x = \sqrt{\frac{\sum x^2}{n} - \left(\frac{\sum x}{n}\right)^2}$$

$\Sigma x^2 = 6750, \Sigma x = 235, \quad n = 10$

$$\sigma_x = \sqrt{\frac{6750}{10} = \left(\frac{235}{10}\right)^2} = \sqrt{675 - 552.25} = 11.08$$

$$\text{S.D} = \text{of } y = \sqrt{\frac{\sum y^2}{n} - \left(\frac{\sum y}{n}\right)^2}$$

$\Sigma y^2 = 6840, \ \ \Sigma y = 250, \ \ n = 10$

$$\sigma_x = \sqrt{\frac{6840}{10} = \left(\frac{250}{10}\right)^2} = \sqrt{685 - 625} = 7.68$$

Normally x is used to denote deviations from mean *i.e.* $x = (X - \overline{X})$ and $\Sigma(x - \overline{X})$ is always zero. But in the given question $\Sigma x = 235$ which means deviation are not taken from actual mean and hence the formula for assumed mean is used.

Example 15:

(a) *Calculate mean deviation from the following series:*

X	*10*	*11*	*12*	*13*	*14*
f	*3*	*12*	*18*	*12*	*3*

Solution:

Calculation of Meandeviation

X	*f*	| *D* |	*f* | *D* |	*c.f.*
10	3	2	6	3
11	12	1	12	15
12	18	0	0	33
13	12	1	12	45
14	3	2	6	48
	N = 48		Σ f | D | = 36	

$$\text{M.D.} = \frac{\sum f\,|D|}{N}$$

$$\text{Median} = \text{Size of } \frac{N+1}{2}\text{th item} = \frac{48+1}{2} = 24.5\text{th item}$$

Size of 24.5th item is 12, hence Median = 12

$$\text{M.D.} = \frac{36}{48} = 0.75.$$

Example 16:

Calculate the mean deviation from the mean for the following data:

Size:	*2*	*4*	*6*	*8*	*10*	*12*	*14*	*16*
Frequency:	*2*	*2*	*4*	*5*	*3*	*2*	*1*	*1*

Solution:

Calculation of Mean Deviation from Mean

X	f	fX	$\lvert X - 8 \rvert$	$f \lvert D \rvert$
2	2	4	6	12
4	2	8	4	8
6	4	24	2	8
8	5	40	0	0
10	3	30	2	6
12	2	24	4	8
14	1	14	6	6
16	1	16	8	8
	N = 20	$\Sigma fX = 160$		$\Sigma f \lvert D \rvert = 56$

$$\overline{X} = \frac{\sum fX}{N} = \frac{160}{20} = 8$$

$$M.D = \frac{\sum f|D|}{N} = \frac{56}{20} = 2.8.$$

Calculation of Mean Deviation-Continuous Series

For calculating mean deviation in continuous series the procedure remains the same as discussed above. The only difference is that here we have to obtain the mid-point of the various classes and take deviations of these points from median. The formaula is same, *i.e.*,

$$M.D. = \frac{\sum f|D|}{N}.$$

Example 17:

Blood serum cholesterol levels of 10 persons are as under:

240, 260, 290, 245, 255, 288, 272, 263, 277, 251.

Calculate standard deviation with the help of assumed mean.

Solution:

Calculation of Standard Deviation by the Assumed Mean Method

x	*(X – 264)** *d*	*d^2*
240	–24	576
260	–4	16
290	+26	676
245	–19	361
255	–9	81
288	+24	576
272	+8	64
263	–1	1
277	+13	169
251	–13	169
$\Sigma X = 2641$	$\Sigma d = +1$	$\Sigma d^2 = 2689$

$$\sigma = \sqrt{\frac{\sum d^2}{N} - \left(\frac{\sum d}{N}\right)^2}$$

$$\Sigma d^2 = 2689, \ \Sigma d = +1, \ \ N = 10$$

$$\sigma = \sqrt{\frac{2689}{10} - \left(\frac{1}{10}\right)^2}$$

$$= \sqrt{268.9 - 0.01} = 16.398$$

Example 18:

Find the median and mean deviation of the following data:

Size	*Frequency*	*Size*	*Frequency*
0-10	*7*	*40-50*	*16*
10-20	*12*	*50-60*	*14*
20-30	*18*	*60-70*	*8*
30-40	*25*		

Solution:

Calculation of Median and Mean Deviation

Size	*f*	*c.f.*	*m.p.* *m*	$\mid m - 35.2 \mid$ $\mid D \mid$	$f \mid D \mid$
0-10	7	7	5	30.2	211.4
10-20	12	19	15	20.2	242.4
20-30	18	37	25	10.2	183.6
30-40	25	62	35	0.2	5.0
40-50	16	78	45	9.8	156.8
50-60	14	92	55	19.8	277.2
60-70	8	100	65	29.8	238.4
	N = 100				$\Sigma f \mid D \mid = 1314.8$

$$\text{Med.} = \text{Size of } \frac{N}{2}\text{th item} = \frac{100}{2}$$

Median lies in the clas 30-40

$$\text{Med.} = L + \frac{N/2 - c.f.}{f} \times i$$

L = 30, N/2 = 50, c.f. = 37, f = 25, i = 10

$$\text{Med.} = 30 + \frac{50 - 37}{25} \times 10 + 30 + 5.2 = 35.2$$

$$\text{M.D.} = \frac{\sum f|D|}{N} = \frac{1314.8}{100} = 13.148$$

MERITS AND LIMITATIONS

Merits. The outstanding advantage of the average deviation is the its relative simplicity. It is simple to understand and easy to compute. Any one familiar with the concept of the average can readily appreciate the meaning of the average deviation. If a situation requires a measure of dispersion that will be presented to the general public or any group not very familiar with in statistics, the average deviation is useful.

(a) Since deviations are taken from a central value, comparison about formation of different distributions can easily be made.

(b) Mean deviation is less affected by the value of extreme itreme items than the standard deviation.

(c) It is based on each and every iterm of the data. Consequently change in the value of any item would change the value of mean deviation.

Limitations. The greatest drawback of this method is that algebraic signs are ignored while taking the deviations of the items. For example if from twenty, fifty is deducted we write 30 and not –30. This is mathematically wrong and makes the method non-algebraic. If the signs of the deviations are not ignored the net sum of deviations will be zero if the reference point is the mean or approximately zero or the reference point is median.

(a) It is rarely used in sociological studies.

(b) It is not capable of further algebraic treatment.

(c) This method may not given us very accurate results. The reason is that mean deviation gives us best results when deviations are taken from median. But median is not a satisfactory measure when the degree of variability in series is very high. And if we compute mean deviation from mean that is also not desirable because the sume of the deviations from mean (ignoring signs) is greater than the sum of the deviations from median (ignoring signs). If mean deviation is computed from mode that is also not scientific because the value of mode cannot always be determined.

Because of these limitations its use is limited and it is overshadowed as a measure of variation by the superior standard deviation.

Usefulness

This measure is useful for small samples with no elaborate analysis required. Incidentally. It may be mentioned that the National Bureau of Economic research has found. in its work on forecasting business cycle, that the average deviation is the most proactical measure of dispersion to use of this purpose.

THE STANDARD DEVIATION

The standard deviation concept was interoduced by Karl Pearson in 1823. It is by far the most important and widely used measure of studying dispersion. Its significance lies in the fact that it is free from those defects from which the earlier methods suffer and satisfies most of the properties of a good measure of dispersion. Standard deviation is also known as *root mean square deviation* for the reason that it is the square root of the mean of the squared deviation from the arithmetic mean. Standard deviation is denoted by the small Greek latter σ (read as sigma).

The standared deviation measures the absolute dispersion (or variability of distribution; the greater the amount of dispersion or variability), the greater the standared deviation, for the greater will be the magnitude of the deviations of the values from their mean. A small standared deviation means a high degree of uniformity of the observation as well as homogeneity of a series: a large standared deviation means just the opposite. Thus if we have two or more comparable series with identical or nearly identical means, it is the distribution with the smallest standard deviation that has the most represenctive means. Hence standared deviation is extremely useful in judgig the representativeness of the mean.

Difference Between Mean Deviation and Standared Deviation

Both these measures of dispersion are based on each and every item of the distribution. but they differ in the following respects:

(a) Mean deviation can be computed either from median or mean. The standard deviation, on the other hand, is always computed from the arithmetic mean because the sum of the squares of the deviation of items from arithmetic mean is the least.

(b) Algebraic signs ae ignored while calculating mean deviation whereas in the calculation of standared deviation signs are taken into account.

Calculation of Standared Deviation

Indiviadual Observations In case of individual observations standard deviation may be computed by applying any of the following two methods:

1. By taking deviation fo the items from the actual mean.
2. By taking deviations of the items from an assumed mean.

Deviations taken from actual mean. When deviations are taken from actual mean the following formula is applied:

$$\sigma * = \sqrt{\frac{\sum x^2}{N}}$$

where, $x = (X - \overline{X})$.

Steps:

(a) Divide $\sum x^2$ by the total number of observations. *i.e.*, N and extract the square-root. This gives us the value of standard deviation.

(b) Square these deviations and obtain the total $\sum x^2$.

(c) Take the deviations of the items from the mean, *i.e.*, fidn $(X - \overline{X})$. Denote these deviations by x.

(d) Calculate the actual mean of the series, *i.e.*, $\overline{X}$.

Deviations taken from Assumed Mean. When the actual mean is in fractions. say, it is 123.674 it would be too caumbersome to take deviations from it and then obtain squares of these deviations. In such a case either the mean may be approximated or else the deivations be taken from an assumed mean and the necessary adjustment made in the value of the standared deviation. The former method of approximation is less accurate and, therefore, invariably in such a case deviations are taken from assumed mean.

When deviations are taken from assumed mean the following formula is applied:

$$\sigma = \sqrt{\frac{\sum fd^2}{N} - \left(\frac{\sum d}{N}\right)^2}$$

(a) substitute the values of $\sum d^2$. $\sum d$ and N in the above formula.

(b) Square these deviations and obtain the total $\sum d^2$.

(c) Take the deviation of the items from an assumed mean, *i.e.*, obtain (X + A). Denote these deviations by d. Take the total of these deviations. *i.e.*, obtain $\sum d$.

Example 19:

Calculate the standard deviation from the following observations:

240.12 240.13240.15240.12240.17

240.15 240.17240.16240.22240.21

Solution:

Calculation of Standard Deviation

X	*(X – 240) d*	d^2
240.12	+0.12	.0144
240.13	+0.13	.0169
240.15	+0.15	.0225
240.12	+0.12	.0144
240.17	+0.17	.0289
240.15	+0.15	.0225
240.17	+0.17	.0289
240.16	+0.16	.0256
240.22	+0.22	.0484
240.21	+0.21	.0441
N = 10	$\sum d = + 1.60$	$\sum d^2 = 0.2666$

$$\sigma = \sqrt{\frac{\sum d^2}{N} - \left(\frac{\sum d}{N}\right)^2} = \sqrt{\frac{2666}{10} - \left(\frac{16}{20}\right)^2} = \sqrt{.02666 - .0256} = 0.035$$

Calculation of Standard Deviation-Discrete Series For calculating standarcd deviation in discrete series any of the following methods may be applied:

1. Actual mean method
2. Assumed mean metho
3. Step deviation method

(a) *Actual Mean Method.* Whn this method is applied, deviations are taken from the actual mean. *i.e.*, we find $(x - \overline{X})$ and denote these deviations by x. These deviations are then squared and multiplied by the respective frequencies. The following formula is applied:

$$\sigma = \sqrt{\frac{\sum fd^2}{N}} \text{ where } x = (X - \overline{X})$$

However, in paractice this method is rarely used because if the actual mean is in fractions the calculations take a tott of time.

(b) *Assumed Mean Method.* When this method is used, the following formula is applied:

$$\sigma = \sqrt{\frac{\sum fd^2}{N} - \left(\frac{\sum fd}{N}\right)^2} \quad \text{where } d = (X - A)$$

(a) Multiply the squared deviations by the respective frequencies, and obtain the total $\sum fd^2$.

(b) Obtain the squares of the deviations, *i.e.*, calculate d^2.

(c) Multiply these deviations by the respective frequencies and obtain the total. $\sum fd$.

(d) Take the deviations of the items from an assumed mean and denote these deviations by d.

Example 20:

The annual salaries of a group of employees are given in the following tabe:

Salaries (in Rs. 000)	*45*	*50*	*55*	*60*	*65*	*70*	*75*	*80*
Number or persons	*3*	*5*	*8*	*7*	*9*	*7*	*4*	*7*

Calculate the standard deviation of the saiaries.

Solution:

Calculation of Standard Deviation

Salaries X	No. of persons f	(X – 60)/5 d	fd	fd^2
45	3	3	–9	27
50	5	–2	–10	20
55	8	–1	–8	8
60	7	0	0	0
65	9	+1	+9	9
70	7	+2	+14	28
75	4	+3	+12	36
80	7	+4	+28	112
	N = 50		$\sum fd = 36 \sum fd2 = 240$	

$$\sigma = \sqrt{\frac{\sum fd^2}{N} - \left(\frac{\sum fd}{N}\right)^2} \times i \; \sum fd^2 = 240,\ N = 50,\ \sum fd = 36,\ i = 5$$

$$\sigma = \sqrt{\frac{240}{50} - \left(\frac{36}{50}\right)^2} \times 5 = \sqrt{4.8 - .5184} \times 5 = 10.35$$

Calculation of Standard Deviation-Continuous Series. In continuous series any of the methods discussed above for discrete frequency distribution can be used. However. in practice it is the step deviation method that is most used. The formula is

$$= \sqrt{\frac{\sum fd^2}{N} - \left(\frac{\sum fd}{N}\right)^2} \times i$$

$$d = \frac{(m - A)}{i} \quad \text{where } i = \text{class interval}$$

Steps.

(a) Square the deviations and multiply them with the respective frequencies of each class and obtain $\sum fd^2$.

(b) Multiply the frequencies of each class with these deviations and obtain $\sum fd$.

(c) Wherever possible take a common factor and denote this column by d.

(d) Take the deviations of these mid-points from an assumed mean and denote these deviations by d.

(e) Find the mid-points of various classes.

Thus the only difference in procedure in case of continuous series is to find mid-points of the various classes.

Example 21:

Calculate mean and standard deviation of following frequency distribution of marks:

Marks	*No. of Students*	*Makes*	*No. of Students*
0.10	*5*	*40-50*	*50*
10-20	*12*	*50-60*	*37*
20-30	*30*	*60-37*	*21*
30-40	*45*		

Solution:

Calculation of Mean and Standard Deviation

Marks	*m.p.* *m*	*f*	*(m – 35)/10* *d*	*fd*	fd^2
0-10	5	5	–3	–15	45
10-20	15	12	–2	–24	48
20-30	25	30	–1	–30	30
30-40	35	45	0	0	0
40-50	45	50	+1	+50	50
50-60	55	37	+2	+74	148
60-70	65	21	+3	+63	189
		N = 200		Σfd = 118	Σfd = 510

$$\overline{X} = A + \frac{\sum fd}{N} \times i = 35 + \frac{118}{200} \times 10 = 35 + 5.9 = 40.9$$

$$\sigma = \sqrt{\frac{\sum fd^2}{N} - \left(\frac{\sum fd}{N}\right)^2} \times 10 = \sqrt{\frac{510}{200} - \left(\frac{118}{200}\right)} \times 10$$

$$= \sqrt{2.55 - 3481} \times 10 = 1.4839 \times 10 = 14.839$$

Example 22:

(a) The mean and S.D of a set of 100 observations were worked out as 40 and 5 respectively. But by mistake a value 50 was taken in place of 40 for one observation, Re-calclulate the correct mean and S.D

Solution:

$$\overline{X} = 40, \sigma = 5, N = 100$$

$$\overline{X} = \frac{\sum X}{N} \Rightarrow 40 = \frac{\sum X}{100}$$

$$\Rightarrow \sum X = 40 \times 100 = 4000$$

$$\text{Correct } \sum X = 4000 - 50 + 40 = 3990$$

$$\text{Correct } \overline{X} = \frac{3990}{100} = 39.9$$

$$\overline{X} = \sqrt{\frac{\sum X^2}{N} - (\overline{X})^2} \Rightarrow \sigma^2 = \frac{\sum X^2}{N} - (\overline{X})^2$$

$$2.5 = \overline{X} = \frac{\sum X^2}{100} - (40)^2$$

$$2500 = \sum X^2 - 160000 \Rightarrow \sum X^2 = 162500$$

$$\text{Correct } \sum X^2 = 162500 - (50)^2 + (40)^2$$

$$= 1625500 - 2500 + 1600 = 161600$$

$$\text{Correct } \sigma = \sqrt{\frac{\text{Correct} \sum X^2}{100} - (\text{Correct } \overline{X})^2}$$

$$= \sqrt{\frac{161600}{100} - (39.9)^2} = \sqrt{1616 - 1592.01} = 4.9.$$

(b) coefficients of variation of two series are 75% and 90% and their standard deviations 15 to 18 respectively. Find their mean.

Solution:

Series A

$$\text{C.V.} = \frac{\sigma}{X} \times 100$$

$$75 = \frac{15}{X} \times 100$$

$$75\overline{X} = 1500 \Rightarrow \overline{X} = 20$$

Series B

$$\text{C.V.} = \frac{\sigma}{X} \times 100$$

$$90 \times \frac{18}{X} \times 100$$

$$90\overline{X} = 1800 \Rightarrow \overline{X} = 20$$

The mean of the two series is the same, *i.e.*, 20.

Example 23:

The scores of two batsmen A and B in ten innings during a certain season are:

A:	*32*	*28*	*47*	*63*	*71*	*39*	*10*	*60*	*96*	*14*
B:	*19*	*31*	*48*	*53*	*67*	*90*	*10*	*62*	*40*	*80*

Find (using coefficient of variation(which of the two batsman, A or B, is more consistent in scoring.

Solution:

In order to find out which of the two batsman is more consistent, we have to compare the coefficient of variation.

X	$(X - \overline{X})$ *x*	x^2	*Y*	$(Y - \overline{Y})$ *y*	y^2
32	–14	196	19	–31	961
28	–18	324	31	–19	361
47	+1	1	48	–2	4
63	+17	289	53	+3	9
71	+25	625	67	+17	289
39	–7	49	90	+40	1600
10	–36	1296	10	–40	1600
60	+14	196	62	+12	144
96	+50	2500	40	–10	100
14	–32	1024	80	+30	900
$\Sigma X = 460$	$\Sigma X = 0$	$\Sigma X^2 = 6500$	$\Sigma Y = 500$	$\Sigma y = 0$	$\Sigma y^2 = 5968$

Batsman A

$$\overline{X} = \frac{\sum X}{N} = \frac{460}{10} = 46$$

$$\sigma = \sqrt{\frac{\sum x^2}{N}} = \sqrt{\frac{6500}{10}} = 25.495$$

$$\text{C.V.} = \frac{25.49}{46} \times 100 = 55.41$$

Batsman B

$$\overline{Y} = \frac{\sum Y}{N} = \frac{500}{10} = 50$$

$$\sigma = \sqrt{\frac{\sum y^2}{N}} = \sqrt{\frac{5968}{10}} = 24.43$$

$$\text{C.V.} = \frac{24.43}{50} \times 100 = 48.86$$

Since coefficient of variation is less in case of batsman B, hence batsman B is more consisent.

Standard deviation of two organisations taken together = Rs. 12.637.

For finding out which orgaination is more equitable in regard to wages, we have to compare the coefficients of variation.

Organisation C

$$C.V. = \frac{\sigma}{X} \times 100$$

$$\sigma = \sqrt{Variance} = \sqrt{100} = 10$$

$$\overline{X} = 60$$

$$\therefore C.V. = \frac{10}{60} \times 100 = 16.67$$

Organisation D

$$C.V. = \frac{\sigma}{X} \times 100$$

$$\sigma = \sqrt{Variance} = \sqrt{144} = 12$$

$$\overline{X} = 48$$

$$\therefore C.V. = \frac{12}{48} \times 100 = 25.$$

Since coefficient of variation in organisation C is less hence it is more equitable in regard to wages

Example 24:

An analysis of monthly wages of workers of two organisations C and D yielded the follwing results:

	Organisation	
	C	***D***
No. of workers	*50*	*60*
Average monthly wages	*Rs. 60*	*Rs. 48*
Variance	*100*	*144*

Obtain the average monthly wage and the standard deviation of wages of all workers in the two organisations taken together. Which organisation is more equitable in regard to wages?

Solution:

$$\overline{X}_{12} = \frac{N_1\overline{X}_1 + N_2\overline{X}_2}{N_1 + N_2}$$

$$N_1 = 50,\ \overline{X}_1 = 60,\ N_2 = 60,\ \overline{X}_2 = 48$$

$$\therefore \overline{X}_{12} = \frac{(50 \times 60) + (60 \times 48)}{50 + 60} = \frac{3000 + 2880}{100} = \frac{5880}{110} = \text{Rs. } 53.45$$

$$\sigma_{12} = \sqrt{\frac{N_1\sigma_1^2 + N_2\sigma_2^2 + N_1d_1^2 + N_2d_2^2}{N_1 + N_2}}$$

$$N_1 = 50,\ \sigma_{12} = 100,\ N_2 = 60,\ \sigma_1^2 = 144$$

$$d_1 - (\overline{X}_1 - \overline{X}_{12}) = (60 - 53.45) = 6.55$$

$$d_2 = (\overline{X}_2 - \overline{X}_{12}) = (48 - 53.45) = -5.45$$

$$\sigma_{12} = \frac{\sqrt{50 \times 100 + 60 \times 144 + 50\,(6.55)^2 + 60(-5.45)^2}}{50 + 60}$$

$$\sigma_{12} = \frac{\sqrt{5000 + 8640 + 2145.125 + 1782.15}}{110}$$

$$= \sqrt{\frac{17567.275}{100}} = \sqrt{159.7025} = 12.637$$

Example 25:

Compete the table showing the frequencies with which words of differant number of the letters occur in the passage given below (omiting punctuation marks) treating as the variable the number of letters in each word, and calculate the coefficient of variation of the distribution. "Statistics are like proposals of marriage—they shold be, they rarely are, studied and considered, very deliberately upon their alround menits."

Solution:

Here the variable (X) is the number of letters in each word. In the given passage there are words with number of letters ranging from 1 to 12. Hence the variable X would take values from 1 to 11. The frequency distribution can easily be formed by using tally bars.

No. of Letters in a word X	*Tally Bars*	*f*	*(X – 6) d*	*fd*	*fd²*
1	–	0	–5	0	0
2	||	2	–4	–8	32
3	||||	5	–3	–15	45
4	||||	4	–2	–8	16
5	||	2	–1	–2	2
6	|||	3	0	0	0
7	|	1	+1	+1	1
8	|	1	+2	+2	4
9	|	1	+3	+3	9
10	||	2	+4	+8	32
11	–	0	+5	0	0
12	|	1	+6	+6	36
		N = 22		$\Sigma fd = -13$	$\Sigma fd^2 = 177$

$$C.V. = \frac{\sigma}{X} \times 100$$

$$\overline{X} = A + \frac{\sum fd}{N} = 6 - \frac{13}{22} = 5.40$$

$$\sigma = \sqrt{\frac{\sum fd^2}{N} - \left(\frac{\sum fd}{N}\right)^2} = \sqrt{\frac{177}{22} - \left(\frac{-13}{22}\right)^2} = \sqrt{8.045 - 0.349} = 2.774$$

$$C.V. = \frac{2.779}{5.409} \times 100 = 51.28 \text{ percent.}$$

Example 26:

Find the mean and standard deviation of first n natural numbers.

Solution:

The first n natural numbers are 1, 2, 3, 4,..., n.

$$\Sigma X = 1 + 2 + 3 + 4 + ... + n = \frac{n(n+1)}{2}$$

$$\Sigma X^2 = (1)^2 + (2)^2 + (3)^2 + ... + n^2$$

$$= \frac{n(n+1)(2n+1)}{6}$$

$$\overline{X} = \frac{\sum X}{n} = \frac{n(n+1)}{2n} = \frac{n+1}{2}$$

$$\sigma^2 = \frac{\sum X^2}{n} - \left(\frac{\sum X}{n}\right)^2$$

$$= \frac{n(n+1)(2n+1)}{6n} - \left\{\frac{n(n+1)}{2n}\right\}^2$$

$$= \frac{(n(n+1)(2n+1)}{6} - \frac{(n+1)^2}{4}$$

$$= \frac{2(n+1)(2n+1) - 3(n+1)^2}{12}$$

$$= \frac{(n+1)(4n + - 3n - 3)}{12}$$

$$= \frac{(n+1)(n-1)}{12} = \frac{n^2 - 1}{12} \Rightarrow \sigma = \sqrt{\frac{n^2 - 1}{12}}$$

Thus the mean of n natural numbers is $\frac{n+1}{2}$ and standard deviation

$$= \sqrt{\frac{n^2 - 1}{2}}$$

Example 27:

The table below gives ths weight measurements of 200 castings:

Weight in Kg.	*No. of Catings*	*Weight in Kg.*	*No. of Castings*
81-90	*2*	*141-150*	*37*
91-100	*5*	*151-160*	*29*
101-110	*13*	*161-170*	*11*
11-120	*20*	*171-180*	*3*
121-130	*30*	*181-190*	*1*
131-140	*49*		

Calculate arithmetic median, mode and standard deviation.

Solution:

Calculation of $\overline{X}$, Med, Mode and S.D.

Weight (Kgs.)	*m.p. m*	*No. of castomgs*	*(m – 125.3)/10 d*	*fd*	*fd²*	*c.f.*
81-90	85.5	2	–4	–8	32	2
91-100	95.5	5	–3	–15	45	7
101-110	105.5	13	–2	–26	52	20
111-120	115.5	20	–1	–20	20	40
121-130	125.5	30	0	0	0	70
131-140	135.5	49	+1	+49	49	119
141-150	145.5	37	+2	+74	148	156
151-160	155.5	29	+3	+87	261	185
161-170	165.5	11	+4	+44	176	196
171-180	175.5	3	+5	+15	75	199
181-190	185.5	1	+6	+6	36	200
		N = 200		Σfd = 206	Σfd = 894	

$$\overline{X} = A + \frac{\sum fd}{N} \times i = 125.5 + \frac{206}{200} \times 10 = 124.5 + 10.3 = 135.8$$

$$\text{Med. Size of } \frac{N}{2}\text{th item} = \frac{200}{2} = 100\text{th item}$$

Median less in the class 131-140, But the real limits of this class is 130.5 - 140.5

$$\text{Med.} = L \ \frac{N/2 - c.f.}{f} \times i = 130.5 + \frac{100 - 70}{49} \times 10$$

$$= 130.5 + 6.12 = 136.62$$

Mode. By inspection mode lies in the class 131-140. The real limits of this class is 130.5 – 140.5.

$$\text{Mode} = L + \frac{\Delta_1}{\Delta_1 + \Delta_2} \times i$$

$$L = 13.5,\ \Delta_1 = (49 - 20) = 19,\ \Delta_2 = (49 - 37) = 12,\ i = 10$$

$$\therefore \text{Mode} = 130 + \frac{19}{19 + 20} \times 10 = 130.5 + 6.13 = 136.63$$

$$\sigma = \sqrt{\frac{\sum fd^2}{N} - \left(\frac{\sum fd}{N}\right)^2} \times 10 \sqrt{\frac{894}{200} - \left(\frac{206}{200}\right)^2} \times 10$$

$$\sqrt{4.471.061} \times 10 = 1.843 \times 10 = 18.46.$$

Example 28:

For two groups of observations the following results were available:

Group I	*Group II*
$\Sigma(X - 5) = 8$	$\Sigma(X - 5) = -10$
$\Sigma(X - 5)^2 = 40$	$\Sigma(X - 8)^2 = 70$
$N_1 = 20$	$N_2 = 25$

Find the mean and the standard deviation of the 45 observations obtained by combining the two groups.

Solution:

Group I	*Group II*
$\Sigma(X - 5) = 8$	$\Sigma(X - 5) = -10$
$\Sigma X - \Sigma 5 = 8$	$\Sigma X - \Sigma 8 = -10$
$\Sigma X - 5 \times 20 = 8$	$\Sigma X - 8 \times 25 = -10$
$\Sigma X = 8 + 100 = 108$	$\Sigma X = -10 + 200 = 190$
$\Sigma(X - 5)^2 = 40$	$\Sigma(X - 8)^2 = 70$

$\Sigma(X^2 - 10X + 25) = 40$ $\Sigma(X^2 - 16X + 64) = 70$

$\Sigma X^2 - 10\,\Sigma X + \Sigma 25 = 40$ $\Sigma X^2 - 16\,\Sigma X + \Sigma 64 = 70$

$\Sigma X^2 - 10 \times 108 + 25 \times 20 = 40$ $\Sigma X^2 - 16 \times 190 + 65 \times 25 = 70$

$\Sigma X^2 - 40 + 1080 - 500 = 620$ $\Sigma X^2 - 70 + 3040 - 1600 = 1510$

For 45 observations in the combined group we have:

$\Sigma X = 108 + 190 = 298$

$\Sigma X^2 = 620 + 1510 = 2130$

Mean of the two groups $= \dfrac{298}{45} = 6.222$

Standard deviation of the two groups

$$\sigma = \sqrt{\frac{\sum X^2}{N} - \left(\frac{\sum X}{N}\right)^2} = \sqrt{\frac{2130}{45} - \left(\frac{298}{45}\right)^2} = \sqrt{47.333 - 42.854}$$

$= 1.865.$

Example 29:

if the values of the mean and standard deviation of the following frequency distribution (obtained by step deviation method) are 135.3 and 9.6 respectively, determine the actual class-intervals:

d	*–4*	*–3*	*–2*	*–1*	*0*	*+1*	*+2*	*+3*	*Total*
t	*2*	*5*	*8*	*18*	*22*	*13*	*8*	*4*	*80*

Solution:

We will first determine the value of i then A and finally find actual class-interval.

d	*f*	*fd*	*fd²*
–4	2	–8	32
–3	5	–15	45
–2	8	–16	32
–1	18	–18	18
0	22	0	0
+1	13	+13	13
+2	8	+16	32
+3	4	+12	36
	N = 80	$\Sigma fd = -16$	$\Sigma fd^2 = 208$

$$\sigma = \sqrt{\frac{\sum fd^2}{N} - \left(\frac{\sum fd}{N}\right)^2} \times i \Rightarrow 9.6 = \sqrt{\frac{208}{80} - \left(\frac{-16}{80}\right)^2} \times i$$

$$= \sqrt{2.6 - 0.04} \times i = 9.6$$

$$\Rightarrow 1.6i = 9.6 \Rightarrow i = 6$$

$$\overline{X} = A + \frac{\sum fd}{N} \times i$$

$$\Rightarrow 135.3 + 1.2 = 136.5$$

The mid-points coresponding to various step deviations shall be

d	–4	–3	–2	–1	0	1	2	3
m.p.	112.5	118.5	124.5	130.5	136.5	142.5	148.5	154.5

The various clase limits shall be obtained as follows:

$$\text{m.p.} \pm \frac{1}{2} \quad \Rightarrow \quad \text{m.p.} = \frac{6}{2}$$

The actual class limits shall be

Class limits	*Frequency*	*Class limits*	*Frequency*
109.5-115.5	2	133.5-139.5	22
115.5-121.5	5	139.5-145.5	13
121.5-127.5	8	145.5-151.5	8
127.7-133.5	18	151.5-157.5	4

Example 30:

The sharehoders Research Centre of India has conducted recently a rsearch-study on price behaviour of three leading industrial shares, A, B and C for the period 1979 to 1985, the results of which are published as follows in its Quarterly Journal:

Share	*Average Price*	*Standard Deviation*	*Current selling Price*
A	*18.2*	*5.4*	*36.00*
B	*22.5*	*4.5*	*34.75*
C	*24.0*	*6.0*	*39.00*

The above figures are given is Rs.

(a) *Which share, in your opinion, appear to be more stable in value?*

(b) *If your are the holder of all the three shares, which one would you like to dispose of at present, and why?*

Solution:

(a) For finding out which share is more stable in value, we have to compare the coefficient of variation of shares A, B and C.

Share A: $\text{C.V.} = \frac{\sigma}{X} \times 100$

$\overline{X} = 18.0, \quad \sigma = 5.4$

$\text{C.V.} = \frac{5.40}{18} \times 100 = 30$

Share B: $\overline{X} = 22.5,$

$\sigma = 4.5$

$\text{C.V.} = \frac{4.0}{22.50} \times 100 = 20$

Share C: $\overline{X} = 24.0,$

$\sigma = 6.0$

$\text{C.V.} = \frac{6.00}{24.00} \times 100 = 25$

Since coefficient of variation is less for share B, hence share B is more stable in value.

(b) I would like to dispose of shares A, since their coefficient of variation is highest, meaning thereby that there is maximum variation in prices.

Example 31:

Calculate the quartile deviation for the data given below:

Daily wages (Rs.)	*No. of wage earners*
35-36	*14*
36-37	*20*
37-38	*42*
38-39	*54*
40-41	*45*
42-42	*21*
42-43	*8*

Solution:

Calculation of Quartile Deviation

Daily wages (Rs.)	*f*	*c.f.*
35-36	14	14
36-37	20	34
37-38	42	76
38-39	54	130
40-41	45	175
41-42	21	196
42-43	8	204

$$\text{Q.D.} = \frac{Q_3 - Q_1}{2}$$

$$Q_1 = \text{Size of } \frac{N}{4}\text{th item} = \frac{204}{4} = 5\text{th item}$$

Q_1 lies in the class 37-38

$$Q_1 = L + \frac{N/4 - c.f.}{f} \times i$$

$$L = 37,\ N/4 = 51,\ c.f. = 34,\ f = 42,\ i = 1$$

$$Q_1 = 37 + \frac{51 - 34}{42} = 37 + 0.405 = 37.405$$

$$Q_3 = \text{Size of } \frac{3N}{4}\text{th item} = \frac{3 \times 204}{4} = 153\text{rd item}$$

Q_3 lies in the class 40-41

$$Q_3 = L + \frac{3N/4 - c.f.}{f} \times i$$

$$L = 40,\ 3N/4 = 153,\ c.f. = 130,\ f = 45,\ i = 1$$

$$Q_3 = 40 + \frac{153 - 130}{45} = 40 + 0.511 = 40.511$$

$$\text{Q.D.} = \frac{40.511 - 37.405}{2} = \frac{3.106}{2} = 1.553$$

Example 32:

An algebra test was given to 400 high school children of whom 150 were and 250 girls. The results were as follows:

$n_1 = 150$ $\qquad$ $n_2 = 250$

$\overline{X}_1 = 72$ $\qquad$ $\overline{X}_2 = 73$

$s_1 = 7.0$ $\qquad$ $s_2 = 6.4$

Find the mean and the standard deviation of combined group.

Solution:

$$\overline{X}_{12} = \frac{n_1\overline{X}_1 + n_2\overline{X}_2}{n_1 + n_2}$$

$n_1 = 150,\ \overline{X}_1 = 72,\ n_2 = 250,\ \overline{X}_2 = 73$

$$\therefore \quad \overline{X}_{12} = \frac{(150 \times 72) + (250 \times 73)}{150 + 250} = \frac{10800 + 18250}{400} = \frac{29050}{400}$$

$$= 72.625$$

$$\sigma_{12} = \sqrt{\frac{n_1\sigma_1^2 + n_2\sigma_2^2 + n_1 d_1^2 + n_2 d_2^2}{n_1 + n_2}}$$

$\sigma_1 = 7,\ \sigma_2 = 6.4,\ d_1 = \left|\overline{X}_1 - \overline{X}_{12}\right| = |72 - 72.625| = 0.6.25$

$d_2 = \left|\overline{X}_2 - \overline{X}_{12}\right| = 73 - 72.625 = 0.375$

$$\sigma_{12} = \sqrt{\frac{150(7)^2 + 250(6.4)^2 + 150(.625)^2 + 250(.375)^2}{400}}$$

$$= \sqrt{\frac{7350 + 10240 + 58.594 + 35.135}{400}} = \sqrt{\frac{17683.7}{400}} = 6.65$$

Thus the combined mean was Rs. 72.63 and standard eviation 6.65.

Example 33:

The mean weight of 150 students is 60 kg. The mean weight of boys is 70 kg with a standard deviation of 10 kg. For the girls, the mean weight is 55 kg and the standard deviation is 15 kg. Find the number of boys and the combined standard deviation.

Solution:

$$\overline{X}_{12} = \frac{N_1\overline{X}_1 + N_2\overline{X}_2}{N_1 + N_2}$$

Given $\overline{X}_{12} = 60, \overline{X}_1 = 70, \overline{X}_2 = 55,$

$$N_1 + N_2 = 150$$

We have to determine the no. of boys

Hence N_2 will be the no. of girls

$$N_2 = (150 - N_1)$$

Putting the values

$$60 = \frac{N_1\ 70 + (150 - N_1)\ 55}{150}$$

$$9000 = 70N_1 + 8250 = 55N_1$$

$$\Rightarrow \quad 15N_1 = 9000 - 8250 = 750$$

$$\Rightarrow \quad N_1 = 750/15 = 50$$

Hence $\quad N_1 = 150 - 50 = 100$

Thus the no. of boys and girls is 50 and 100 respectively.

Combined Standard Deviation

$$\sigma_{12} = \sqrt{\frac{N_1\sigma_1^2 + N_2\sigma_2^2 + N_1d_1^2 + N_2d_2^2}{N_1 + N_2}}$$

$$N_1 = 50,\ \sigma_1 = 10,\ N_2 = 100,\ \sigma_2 = 15$$

$$d_1 = \left|\overline{X}_1 - \overline{X}_{12}\right| = 60 - 70 = 10,\ d_2 = \left|\overline{X}_2 - \overline{X}_{12}\right| = |55{-}60| = 5$$

$$\sigma_{12} = \sqrt{\frac{50(10)^2 + 100(15)^2 + 50(10)^2 + 100(5)^2}{50 + 100}}$$

$$\sigma_{12} = \sqrt{\frac{5000 + 22500 + 5000 + 2500}{150}} = \sqrt{\frac{35{,}000}{150}} = 15.28.$$

Example 34:

Calculate coefficient of quartile deviation and coefficient of variation from the following data:

Maks	*N. of Students*
Below 20	*8*
" *40*	*20*
" *60*	*50*
" *80*	*70*
" *100*	*80*

Solution:

Calculation of Coeff. of Q.D. and Coeff. of Variation

Marks	*m.p.* *m*	*No. of Students*	*(m – 50)/20* *d*	*fd*	*fd²*	*c.f*
0-20	10	8	–2	–16	32	8
20-40	30	12	–1	–12	12	20
40-60	50	30	0	0	0	50
60-80	70	20	+1	+20	20	70
80-100	90	10	+2	+20	40	90
		N = 80		$\sum fd = 12$	$\sum fd^2 = 104$	

$$C.V. = \frac{\sigma}{X} \times 100$$

$$\overline{X} = A \frac{\sum fd}{N} \times i = 50 + \frac{12}{80} \times 20 = 50 + 3 = 53$$

$$\sigma = \sqrt{\frac{\sum fd^2}{N} - \left(\frac{\sum fd}{N}\right)^2} \times i = \sqrt{\frac{104}{80} - \left(\frac{12}{80}\right)^2} \times 20$$

$$= \sqrt{1.3 - 0.225} \times 20 = 1.1303 \times 20 = 22.606$$

$$C.V = \frac{22.606}{53} \times 100 = 42.65 \text{ percent}$$

$$\text{Coeff. of Q.D.} = \frac{Q_3 - Q_1}{Q_3 + Q_1}$$

$$Q_1 = \text{Size of } \frac{N}{4}\text{th item} = \frac{80}{4} = 20\text{th item}$$

Q_1 lies in the class 20-40

$$Q_1 = L + \frac{N/4 - c.f.}{f} \times i = 20 + \frac{20 - 8}{12} \times 20 = 20 + 20 = 40$$

$$Q_3 = \text{Size of } \frac{3N}{4}\text{th item} = \frac{3(8)}{4} = 60\text{th item}$$

Q_3 lies in the class 60-80

$$Q_3 = L + \frac{3N/4 - c.f.}{f} \times i = 60 + \frac{60 - 50}{20} \times 20 = 60 + 10 = 70$$

$$\text{Coeff. of Q.D.} = \frac{70 - 40}{70 + 40} = \frac{30}{110} = 0.273.$$

Example 35:

Calculate the standared deviation from the data given below:

Size of item	*Frequency*	*Size of item*	*Frequency*
3.5	*3*	*7.5*	*85*
4.5	*7*	*8.5*	*32*
5.5	*22*	*9.5*	*8*
6.5	*60*		

Solution:

Calculation of Standard Deviation

X *Size of item*	*f*	*(X – 6.5)* *d*	*fd*	*fd²*
3.5	3	–3	–9	27
4.5	7	–2	–14	28
5.5	22	–1	–22	22
6.5	60	0	0	0
7.5	85	+1	+85	85
8.5	32	+2	+64	128
9.5	8	+3	+24	72
	N = 227		Σfd = 128	Σfd^2 = 362

$$\sigma = \sqrt{\frac{\sum fd^2}{N} - \left(\frac{\sum fd}{N}\right)^2}$$

$\Sigma fd^2 = 362$, $\Sigma fd = 128$, $N = 217$

$$\sigma = \sqrt{\frac{362}{217} - \left(\frac{128}{217}\right)^2}$$

$$= \sqrt{1.668 - .348} = 1.149$$

(c) *Step Deviation Method.* When this method is used we take deviations of midpoints from an assumed mean and divide these deviations by the with of lass interval, *i.e.*, 'i'. In case class intervals are unequal, we divide the deviatons of midpoints by the lowest common factor and use 'c' instead of 'i' in the formula for calculating standard deviation. The formula for calculating standard deviation is:

$$\sigma = \sqrt{\frac{\sum fd^2}{N} - \left(\frac{\sum fd}{N}\right)^2} \times i$$

where, $d = \frac{(X - A)}{i}$ and i = class interval.

The use of the above formula simplifies calculations.

Example 36:

Find the standard deviation from the following data:

Age under:	*10*	*20*	*30*	*40*	*50*	*60*	*70*	*80*
No. of person dying:	*15*	*30*	*53*	*75*	*100*	*110*	*115*	*125*

Solution:

Calculation of Standard Deviation

Marks	**f** **m**	**m.p.** **m**	**(m – 35)/10** **d**	**fd**	**fd²**
0-10	15	5	–3	–45	135
10-20	15	15	–2	–30	60
20-30	23	25	–1	–23	23
30-40	22	35	0	0	0
40-5ᴗ	25	45	+1	+25	25
50-60	10	55	+2	+20	40
60-70	5	65	+3	+15	45
70-80	10	75	+4	+40	160
	N = 125			Σfd = 2	Σfd² = 488

$$\sigma = \sqrt{\frac{\sum fd^2}{N} - \left(\frac{\sum fd}{N}\right)^2} \times i = \sqrt{\frac{448}{125} \times \left(\frac{2}{125}\right)^2} \times 10$$

$$= \sqrt{3.904 - 0003} \times 10 = 1.976 \times 10 = 19.76.$$

Example 37:

The tollowing are some of the particulars of the distribution of weight of boys and giris in a class

	Boys	*Girls*
Number	*100*	*50*
Mean weight	*60 kg*	*45 kg*
Variance	*9*	*4*

(a) Find the standard deviation of the combined data

(b) Which of the two distributions is more variable?

Solution:

(a) Combined S.D. $\sigma_{12} = \sqrt{\dfrac{N_1\sigma_1^2 + N_2\sigma_2^2 + N_1\sigma_1^2 + N_2\sigma_2^2}{N_1 + N_2}}$

For finding combined standard deviation, we have to calculate combined mean.

$$\overline{X}_{12} = \frac{N_1\overline{X}_1 + N_2\overline{X}_2}{N_1 + N_2}$$

$$= \frac{100\ (60) + 50\ (45)}{100 + 50} = \frac{6000 + 2250}{150} = 55$$

$N_1 = 100,\ \sigma_1^2 = 9,\ N_2 = 50,\ \sigma_2^2 = 4,$

$d_1 = |\overline{X}_1 - \overline{X}_{12}| = 60 - 55 = 5$

$d_2 =- |\overline{X}_2 - \overline{X}_{12}| = |45 - 55| = 10.$

Substituting the values

$$\sigma_{12} = \sqrt{\frac{100\ (9) + 50\ (4) + 100\ (5)^2 + 50\ (10)^2}{150}}$$

$$= = \sqrt{\frac{900 + 200 + 2500 + 5000}{150}} = \sqrt{\frac{8600}{150}} = 7.57$$

(b) For finding which distribution is more variable compare the coefficient of variation of two distributions:

$$\text{C.V. (Boys)} = \frac{\sigma}{X} \times 100 = \frac{3}{60} \times 100 = 5.00$$

$$\text{C.V. (Girls)} = \frac{\sigma}{X} \times 100 = \frac{2}{45} \times 100 = 4.44$$

Since coefficint of variation is more for distribution of weight of boys hence this distribution shows greater variability.

Example 38:

Find the standard deviation of the following distribution:

Age:	*20-25*	*25-30*	*30-35*	*35-40*	*40-45*	*45-50*
No. of persons:	*170*	*110*	*30*	*45*	*40*	*35*

Take assumed average = 32.5.

Solution:

Calculation of Standard Deviation

Age	*m.p.* *m*	*No. of persons* *f*	*(m-32.5)/5* *d*	*fd*	*fd²*
20-25	22.5	170	–2	–340	680
25-30	27.5	110	–1	–110	110
30-35	32.5	80	0	0	0
35-40	37.5	45	+1	+45	45
40-45	32.5	40	+2	+80	160
45-50	47.5	35	+3	+105	315
		N = 480		Σfd = – 220	Σfd² = 1310

$$\sigma = \sqrt{\frac{\sum fd^2}{N} 0 \left(\frac{\sum fd}{N}\right)^2} \times i = \sqrt{\frac{1310}{480} - \left(\frac{-220}{480}\right)^2} \times 5$$

$$= \sqrt{2.729 - 21} \times 5 = \sqrt{2.519} \times 5 = 1.587 \times 5 = 7.936.$$

Methematical Properties of Standard Deviation

Standard deviation has some very importanı mathematical properties which considerably enhance its utility in statistical work.

1. Combined Standard Deviation

Just as it is possible to compute combined mean of two or more than two groups, similarly we can also compute combined standard deviation of two or more groups. Combined standared deviation is denoted by σ_{12} and is computed as follows:

$$\sigma_{12} = \sqrt{\frac{N_1\sigma_1^2 + N_2\sigma_2^2 + N_1\sigma_1^2 + N_2\sigma_2^2}{N_1 + N_2}}$$

σ_2 = combined standard deviation;

σ_1 = standard deviation of first group;

σ_2 = standard deviation of second group;

$d_1 = (\overline{X}_1 - \overline{X}_{12})$; $d_2 = (\overline{X}_2 - \overline{X}_{12})$

The above formula can be extended to find out the standard deviation of three or more groups. For example, combined standard deviation of three groups would be:

$$\sigma_{123} = \sqrt{\frac{N_1\sigma_1^2 + N_2\sigma_2^2 + N_3\sigma_3^2 + N_1d_1^2 + N_2d_2^2 + N_3d_3^2}{N_1 + N_2 + N_3}}$$

where $d_1 = (\overline{X}_1 - X_{123})$; $d_2 = (\overline{X}_2 - X_{123})$; $d_3 = (\overline{X}_3 - X_{123})$.

Example 39:

The number of workers employed, the mean wage (in Rs.) per month and standared deviation (in Rs.) in each section of a factory are given below. Calculate the mean wages and standard deviation of all the workers taken together.

Section	***No. of workers employed***	***Mean wages in Rs.***	***Standard deviation in Rs.***
A	*50*	*1113*	*60*
B	*60*	*1120*	*70*
C	*90*	*1115*	*80*

Solution:

$$\overline{X}_{123} = \frac{N_1\overline{X}_1 + N_2\overline{X}_2 + N_3\overline{X}_3}{N_1 + N_2 + N_3}$$

$$= \frac{(50 \times 1113) + (60 \times 1120) + (90 \times 1115)}{50 + 60 + 90}$$

$$= \frac{55{,}650 + 67{,}200 + 1{,}00{,}350}{200}$$

$$= \frac{2,23,200}{200} = \text{Rs. } 1116.$$

Combined standard deviation of three series:

$$\sigma_{123} = \sqrt{\frac{N_1\sigma_1^2 + N_2\sigma_2^2 + N_3\sigma_3^2 + N_1\sigma_1^2 + N_2\sigma_2^2 + N_3\sigma_3^2}{N_1 + N_2 + N_3}}$$

$$d_1 = |\overline{X}_1 - \overline{X}_{123}| \text{ or } |1113 - 1116| = 3$$

$$d_2 = |\overline{X}_2 - \overline{X}_{123}| \text{ or } |1120 - 1116| = 4$$

$$d_3 = |\overline{X}_3 - \overline{X}_{123}| \text{ or } |1115 - 1116| = 1$$

$$\sigma_{123} = \sqrt{\frac{50(60)^2 + 60(70)^2 + 0(80)^2 + 50(3)^2 + 60(4)^2 + 90(1)^2}{50 + 60 + 90}}$$

$$= \sqrt{\frac{1,80,000 + 2,94,000 + 5,76,000 + 450 + 960 + 90}{200}}$$

$$= \sqrt{\frac{10,51,500}{200}} = \sqrt{5.257.5} = 72.51$$

Standard Deviation of n Natural Numbers. The standard deviation of the first n natural numbers* can be obtained by the following formula:

$$\sigma = \sqrt{\frac{1}{12}(N^2 - 1)}$$

Thus the standard deviation of natural numbers 1 to 10 will be

$$\sigma = \sqrt{\frac{1}{12}(10^2 - 1)} = \sqrt{\frac{1}{12} \times 99} = \sqrt{8.25} = 2.87.$$

Note: The answer would be the same when direct method of ealculating standard deviation used. But this holds good only for natural numbers.

The Sum of the Squares of the Deviations of Items in the Series from their Arithmetic is Minimum. In order words, the sum of the squares of the deviations of items of any series from a value other than the arithmetic mean would always be greater. This is the why standard deviation is always computed from the arithmetic mean.

The *Standard Deviation Enables us to Determine, with a Great Deal of accuracy, where the Values of a Frequency Distribution are Located.* With the help of Tchebycheffs theorem given by mathematician P.L. Tchebycheff (1821-1894), no matter what the shape of the distribution is, at least 75 percent of the values will fall within ±2 standard deviations from the mean of the distribution, and at least 89 percent of the values will be within ±3 standard deviations from the mean. With the help of normal curve we can measure even with greater precision the number of items the fall within specific ranges.

For a symmetrical distribution, the following relationships hold good:

Mean $\pm 1\ \sigma$ covers 68.27% of the items.

Mean $\pm 2\ \sigma$ covers 95.45% of the items.

Mean $\pm 3\ \sigma$ covers 99.73% of the items.

This can be illustrated by the following diagram:

Relation Between Measures of Dispersion : In a normal distribution there is a fixed relationship between the three most commonly used measures of dispersion. The quartile deviation is smallest, the mean deviation next and the standard deviation is largest, in the following proportions :

$$\text{Q.D.} = \frac{2}{3}\sigma \text{ or } \sigma = \frac{3}{2}\text{Q.D. and M.D.} = \frac{4}{5}\sigma \text{ or } \sigma = \frac{5}{4}\text{ M.D.}$$

These relationships can be easily memorized because of the sequence 2. 3. 4. 5. The same properties tend to hold true for many distributions that are *quite normal*. They are useful in eastimating one measure of dispersion when by natural numbers we mean only positive integers, e.g., 1, 2, 3, 4, 5, ... n. Another is known or in checking roughly the accuracy of a calculated value. If the computed σ differs very widely from its value estimated from Q.D. or M.D. either an error has been made or the distribution differes considerably from normal.

Another comparison may be made of the proportion of items that are typically included within the range of one Q.D., M.D. or S.D. measured both above and below the mean. In a normal distribution.

$\overline{X} \pm$ Q.D. includes 50 per cent of the items.

$\overline{X} \pm$ M.D. includes 57.31 per cent of the items

$\overline{X} \pm \sigma$ includes 68.27 per cent or about two-thirds of items.

Example 40:

An analysis of the monthly wages paid to workers in two firms A and B, belonging to the same industry, gives the following result :

	Firm A	***Firm B***
Number of wage earners	550	650
Average monthly wages	Rs. 1450	Rs. 1400

Standard deviation of the distribution of wagesRs. $\sqrt{10{,}000}$ Rs. $\sqrt{19{,}600}$

Answer the following questions with proper justifications :

(a) *Which firm, or B, pays out larger amount as weekly wages?*

(b) *In which firm, A or B, is there greater variability in individual wages?*

(c) *What are the measures of (i) average weekly wages and (ii) standard deviation of individual wages of all workers in the two firms taken together?*

Solution:

In order to find out which firm A or B pays larger amount of weekly wages, we compare the total wages bill of firm A and firm B.

Firm A Total wage bill = 550 × 1450 = Rs. 7,97,500

Firm B Total wage bill = 650 × 1400 = Rs. 9,10,000

(b) To determine the firm in which there is greater variability in individual wages, we shall compare the coefficient of variation.

Firm A $\text{C.V.} = \frac{\sigma}{\overline{\overline{X}}} \times \frac{\sqrt{10{,}000}}{1450} \times 100 = 6.89$ per cent

Firm B $\text{C.V.} = \frac{\sigma}{\overline{\overline{X}}} \times 100 = \frac{\sqrt{19{,}600}}{1400} \times 100 = 10$ per cent.

Since coefficient of variation is more in case of firm B, hence there is greater variation in the distribution of wages of firm B.

(c) Combined Mean and Standard Deviation

$$\overline{X}_{12} = \frac{N_1 \overline{X}_1 + N_2 \overline{X}_2}{N_1 + N_2}$$

$$= \frac{(550 \times 1450) + (650 \times 1400)}{550 + 650} = \frac{7{,}97{,}500 + 9{,}10{,}000}{1200} = \text{Rs. } 1422.9$$

$$\sigma_{12} = \sqrt{\frac{N_1\sigma_1^2 + N_2\sigma_2^2 + N_1 d_1^2 + N_2 d_2^2}{N_1 + N_2}}$$

$$d_1 = \left|\overline{X}_1 - \overline{X}_{12}\right| = \left|1450 - 1422.92\right| = 27.08$$

$$d_2 = \left|\overline{X}_2 - \overline{X}_{12}\right| = \left|1400 - 1422.92\right| = -22.92$$

$$\sigma_{12} = \sqrt{\frac{(550 \times 10{,}000) + (650 \times 19{,}600) + 550\,(27.08)^2 + 650\,(-22.92)^2}{1200}}$$

$$= \sqrt{\frac{5,00,000 + 1,27,40,000 + 4,03,329.52 + 3,41,462.16}{1200}}$$

$$= \sqrt{\frac{1,89,89,791.68}{1200}} = 39.78$$

Variance

The term variance was used to describe the square of the standard deviation by R.A. Fisher in 1913. The concept of variance is highly important in advanced work where it is possible to split the total into several parts, each attributable to one of the factors causing variation in their original series. Variance is defined as follows :

$$= \sqrt{\frac{18984791.68}{1200}} = \sqrt{15820.66} = 125.78$$

$$\text{Variance} = \frac{\Sigma\left(X - \overline{X}\right)^2}{N}$$

For details please refer to chapter on 'Analysis of Variance'.

Thus, variance is nothing but the square of the standard deviation

i.e. Variance $= \sigma^2$

$\Rightarrow$ $\sigma = \sqrt{\text{Variance}}$

In a frequency distribution where deviations are taken from assumed mean variance may directly be computed as follows:

$$\text{Variance} = \left\{\frac{\Sigma\, fd^2}{N} - \left(\frac{\Sigma\, fd}{N}\right)^2\right\} \times i^2$$

when $d = \frac{(X - A)}{i}$ and i = common factor.

VARIANCE AND STANDARD DEVIATION COMPARED

Both the variance and the standard deviation are measures of variability in a population. These two measures are closely related as is clear from the formula : Variance = σ^2. Variance is the average squared deviation from the arithmetic mean and standard deviation is the square root of the variance. In a subsequent chapter the significance of variance analysis will be discussed at length. The smaller the value of σ^2 the lesser the variability or greater the uniformity in the population.

Example 41:

The following table gives the length of life of 400 radio tubes:

Langth of life (hours) No. of Radio tubes Length of life(hours) No. of Radio tubes

1,000-1, 199	*12*	*2,000-2,199*	*55*
1,200-1,399	*30*	*2,200-2,399*	*36*
1,400-1599	*65*	*2,400-2,599*	*25*
1,600-1,799	*78*	*2,600-2,799*	*9*
1,800-1,999	*90*		

Calculate (i) the average length of life of a radio tube, (ii) the standard deviation of the length of a tube, and (iii) the percentage number of tubes where length of ife of a tube falls within $\overline{X} \pm 2\sigma$.

Solution:

Calculation of Mean and Standard Deviation

Length of life (hours)	*m.p. m*	*No. of Radio tubes f*	*(m – 1899.5)/200 d*	*fd*	*fd²*
1000-1199	1099.5	12	–4	–48	192
1200-1399	1299.5	30	–3	–90	270
1400-1599	1499.5	65	–2	–130	260
1600-1799	1699.5	78	–1	–78	78
1800-1999	1899.5	90	0	0	0
2000-2199	2099.5	55	+1	+55	55
2200-2399	2299.5	36	+2	+72	144
2400-2599	2499.5	25	+3	+75	225
2600-2799	2699.5	9	+4	+36	144
		N = 400	+1	Σfd = 108	Σfd^2 = 1368

(i) $\overline{X} = A + \frac{\sum fd}{M} \times i\ 1899.5 - \frac{108}{400} \times 200 = 1899.5 - 54 = 1845.5$

(ii) $\sigma = \sqrt{\frac{\sum fd^2}{N} - \left(\frac{\sum fd}{N}\right)^2} \times i = \sqrt{\frac{1368}{400} - \left(\frac{-108}{400}\right)^2} \times 200$

$$= \sqrt{3.42 - 0.0\,73} \times 200 = \sqrt{3.347} \times 200 = 1.829 \times 200 = 365.8$$

(iii) $\overline{X} \pm 2\sigma = 1845.5 \pm 2\ (365.8) = 1113.9$ to 2577.1.

We have to determine the percentage of tubes lying between 1113.9 and 2577.1. For this we make an assumption that the litems are equally distributed within each class. In the class 1000-1199 there are 12 frequencies. At 1113.9 or 1114 there would be 6.84 frequencies $\left(\frac{12}{200} \times 114 = 6.84\right)$. Frequencies between 1114 and 1199 are 12-6.84 = 5.16. Between 2400 and 2599, there are 25 frequencies. At 2577.1 or 2577 there would be 22.1 frequencies $\left(\frac{25}{200} \times 117 = 22.1\right)$

Thus the total frequencies between 1113.9 and 2577.1 would be 5.16 + 30 + 65 + 78 + 90 + 55 + 36 + 22.1 = 381.26 or $\frac{381.26}{400} \times 100 = 95.32$ percent.

Coefficient of Variation

The standard deviation discussed above is an absolute measure of dispersion. The corresponding relative measure is known as the *coefficient of variation.* This measure developed by Karl Pearson is the most commonly used measure of relative variation. It is used in such problems where we want to compare the variability of two or more than two series. That series (or group) for which the coefficient of varition is greater is said to be more variable or conversely less consistent, less uniform, less stable or less homogeneous. On the other hand, the series for which coefficient of variation is less is said to be less variable or more consistent, more uniform, more stable or more homogeneous. Coefficient of varation is denoted by C.V. and is obtained as follows:

$$\text{Coefficient of variation or C.V.} = \frac{\sigma}{\overline{X}} \times 100$$

It may be pointed out that although any measure of dispersion can be used in conjunction with any average in computing relative dispersion, statisticians, in fact, almost always use the standared deviation as the measure of dispersion and the arithmetic mean as the average. When the relative dispersion is stated in terms of the arithmetic mean and the standard deviation, the resulting percentage is known as the coefficient of variation or coefficient of variation or coefficient of variability.

Example 42:

Goals scored by two teams in a Football session were as follows:

No. of Goals Scored in a Football Match	*No. of Football Matches Played* *Team 'A'*	*Team 'B'*
0	*15*	*20*
1	*10*	*10*
2	*07*	*05*
3	*05*	*04*
4	*03*	*02*
5	*02*	*01*
Total	*42*	*42*

Calculate coefficient of variation and state which team is more consistent.

Solution:

In order to find out which team is more consistent we shall have to compare the coefficient of variation.

X	**(X – 7)** **x**	**x^2**	**Y**	**(Y – 7)** **y**	**y^2**
15	+8	64	20	+13	169
10	+3	9	10	+3	9
7	0	0	5	–2	4
5	–2	4	4	–3	9
3	–4	16	2	–5	25
2	–5	25	1	–6	36
$\Sigma X = 42$	$\Sigma x = 0$	$\Sigma x^2 = 118$	$\Sigma Y = 42$	$\Sigma y = 0$	$\Sigma y^2 = 252$

Team A

$$C.V. = \frac{\sigma}{X} \times 100$$

$$X = \frac{\sum X}{N} = \frac{42}{6} = 7$$

$$\sigma = \sqrt{\frac{\sum x^2}{N}} = \sqrt{\frac{118}{6}} = 4.43$$

Team B

$$C.V. = \frac{\sigma}{Y} \times 100$$

$$Y = \frac{\sum X}{N} = \frac{42}{6} = 7$$

$$\sigma = \sqrt{\frac{\sum y^2}{N}} = \sqrt{\frac{282}{6}} = 6.48$$

$$\text{C.V.} = \frac{4.43}{7} \times 100 = 63.29 \qquad \text{C.V.} = \frac{6.48}{7} \times 100 = 92.57$$

Example 43:

From the prices of shares of X and Y below find out which is more stable in value:

X	*35*	*54*	*52*	*53*	*56*	*58*	*52*	*50*	*51*	*49*
Y	*108*	*107*	*105*	*105*	*106*	*107*	*104*	*103*	*104*	*101*

Solution:

In order to find out which shares are more stable, we have to compare coefficient of variations.

Calculation of Coefficient of Variation

X	**$(X-\overline{X})$** **x**	**x^2**	**Y**	**$(Y-\overline{Y})$** **y**	**y^2**
35	–16	25	108	+3	9
54	+3	9	107	+2	4
52	+1	1	105	0	0
53	+2	4	105	0	0
56	+5	25	106	+1	1
58	+7	49	107	+2	4
52	+1	1	104	–1	1
50	–1	1	103	–2	4
51	0	0	104	–1	1
49	–2	4	101	–4	16
$\Sigma X = 510$	$\Sigma x = 0$	$\Sigma x^2 = 350$	$\Sigma Y = 1050$	$\Sigma y = 0$	$\Sigma y^2 = 40$

Coefficient of Variation X:

$$\text{C.V.} = \frac{\sigma}{X} \times 100$$

$$\overline{x} = \frac{\Sigma x}{N} = \frac{510}{10} = 51$$

$$\sigma = \sqrt{\frac{\Sigma x^2}{N}} = \sqrt{\frac{350}{10}} = 5.916$$

$$\text{C.V.} = \frac{5.916}{51} \times 100 = 11.6$$

Coefficient of Variation Y:

$$\text{C.V.} = \frac{\sigma}{\text{Y}} \times 100$$

$$\overline{\text{Y}} = \frac{\Sigma \text{Y}}{\text{N}} = \frac{1050}{10} = 105$$

$$\sigma = \sqrt{\frac{\Sigma y^2}{\text{N}}} = \sqrt{\frac{40}{10}} = 2$$

$$\text{C.F.} = \frac{2}{105} \times 100 = 1.905$$

Since coefficient of variation is much less in case of shares Y, hence they are more stable in value.

MERITS AND LIMITATIONS

Merits

The standard deviation is the best measure of variation because of its mathematical characteristics. It is based on every item of the distribution. Also it is amenable to algebraic treatment and is less affected by fluctuations of sampling than most other measures of dispersion.

(a) It is possible to calculate the combined standard deviation of two or more groups. This not possible with any other measure.

(b) For comparing the variability of two or more distributions coefficient of variation is considered to be most appropriate and this is based on mean and standard deviation.

(c) Standard deviation is most prominently used in further statistical work. For example, in computing skewness, correlation, etc. use is made of standard deviation. It is keynote in sampling and provides a unit of measurement for the normal distribution.

Limitations

As compared to other measures it is difficult to compute. However, this does not reduce the importance of this measure because of high degree of accuracy of results it gives.

(a) It gives more weight to extreme items and less to those which are near the mean. It is because of the fact that the squares of the deviations which are big in size would be proportionately greater than the squares of those deviations which are comparatively small. The deviations 2 and 8 are in the ratio of 1 : 4 but their squares, *i.e.* 4 and 64, would be in the ratio of 1 : 16.

CORRECTING INCORRECT VALUES OF MEAN AND STANDARD DEVIATION

Mistakes in calculations are always possible. Sometimes it so happens that while calculating mean and standard deviation we unconsciously copy out wrong items. For example, an item 21 may be copied as 12. Similarly, one item 127 may be taken as only 27. In such cases if the entire calculations are done again, it would become too tedeious a task. By adopting a very simple procedure we can correct the incorrect values of mean and standard deviation. For obtaining correct mean we find out correct ΣX by deducting from the original ΣX the wrong item and adding to it the correct item. Similarly for calculating correct standard deviation we obtain the value of correct ΣX^2. The following illustrations shall clarify the calculations.

Example 44:

The number of employees, wages per employee and the variance of the wages per employee for two factories is given below:

	Factory A	***Factory B***
Number of employees	*100*	*150*
Average per employee per month (Rs.)	*3200*	*2800*
Variance of the wages per employee per month (Rs.)	*625*	*729*

(a) In which factory is there greater variation in the distribution of wages per employee?

(b) Suppose in factory B, the wages of an employee were wrongly noted as Rs. 3050 instead of Rs. 3650, what would be the correct variance for factory B?

Solution:

(a) For finding out in which factory there is greater variation we compare the coefficient of variation.

Factory A

$$\text{C.V.} = \frac{\sigma}{\overline{X}} \times 100$$

$$\sigma = \sqrt{625} = 25, \overline{X} = 3200$$

$$\text{C.V.} = \frac{25}{3200} \times 100 = 0.781$$

Factory B

$$\text{C.V.} = \frac{\sigma}{\overline{X}} \times 100$$

$$\sigma = \sqrt{729} = 27, \overline{X} = 2800$$

$$\text{C.V.} = \frac{27}{2800} \times 100 = 0.964$$

Since coefficient of variation is more in factory B, there is greater variation in the distribution of wages per employee.

(b) **Correct variance** For finding out correct variance, we have to find out the correct mean.

$$\overline{X} = \frac{\Sigma X}{N}, \; N\overline{X} = \Sigma X = 150 \times 2800 = 420000$$

$$\text{Correct } \Sigma X = 420000 - 3050 + 3650 = 420600$$

$$\text{Correct } \overline{X} = \frac{420600}{150} = 2804$$

$$\text{Variance} = \frac{\Sigma X^2}{N} - \left(\overline{X}\right)^2$$

$$729 = \frac{\Sigma X^2}{150} - (2800)^2$$

$$109350 = \Sigma x^2 - 11760000$$

$$\Sigma x^2 = 1176109350$$

$$\text{Correct } \Sigma x^2 = 1176109350 - (3050)^2 + (3650)^2$$

$$= 1176109350 - 9302500 = 1180129350$$

$$\text{Correct variance} = \frac{\text{Correct } \Sigma X^2}{N} - \left(\text{Correct } \overline{X}\right)^2$$

$$= \frac{1180129350}{150} - (2804)^2$$

$$= 7867529 - 7862416 = 5113$$

Example 45:

The following table shows that monthly expenditures of 80 students of a University on morning breakfast:

Expenditure (in Rs.)	***No. of Students***	***Expenditure (in Rs.)***	***No. of Students***
78 – 82	*2*	*53 – 57*	*13*
73 – 77	*6*	*48 – 52*	*9*
68 – 72	*7*	*43 – 47*	*7*
63 – 67	*12*	*38 – 42*	*4*
58 – 62	*18*	*33 – 37*	*2*

Calculate arithmetic mean, standard devaition and coefficient of variation of the above data.

Solution:

Calculation of $\overline{X}$ S.D. and C.V.

Expenditure (Rs.)	m.p. m	f	(m – 60)/5 d	fd	fd²
78 – 82	80	2	+4	+8	32
73 – 77	75	6	+3	+18	54
68 – 72	70	7	+2	+14	28
63 – 67	65	12	+1	+12	12
58 – 62	60	18	0	0	0
53 – 57	55	13	–1	–13	13
48 – 52	50	9	–2	–18	36
43 – 47	45	7	–3	–21	63
38 – 42	40	4	–4	–16	64
33 – 37	35	2	–5	–10	50
		N = 80		Σfd = – 26	Σ fd² = 352

Mean : $\overline{X} = A + \frac{\Sigma fd}{N} \times i = 60 - \frac{26}{80} \times 5 = 60 - 1.625 = 58.375$

S.D. : $\sigma = \sqrt{\frac{\Sigma fd^2}{N} - \left(\frac{\Sigma fd^2}{N}\right)^2} \times i = \sqrt{\frac{252}{80} - \left(\frac{-26}{80}\right)^2} \times 5$

$= \sqrt{4.4 - .106 \times 5 = 2.072 \times 5 = 10.36}$

C.V. $= \frac{\sigma}{\overline{X}} \times 100 = \frac{10.36}{58.375} \times 100 = 17.75\%.$

Example 46:

The first of two sub-groups has 10 items with mean 15 and standard deviation 3. If the whole group has 250 items with mean 15.6 and standard deviation $\sqrt{13.44}$, find the standard deviation of the second sub-group.

Solution:

$N_1 = 100, \ \overline{X}_1 = 15, \ \sigma_1 = 3$

$N_1 + N_2 = 250, \ \overline{X}_2 = 15, \ \sigma_{13} = \sqrt{13.44}$

Since $N_1 + N_2 = (250 - 100) = 150,$

$\Rightarrow \quad 15.6 = \frac{100 \times 15 + 150\overline{X}_2}{250}$

$$3900 = 1500 + 150\overline{X}_2$$

$$150\overline{X}_2 = 2{,}400 \Rightarrow \overline{X}_2 = 16$$

$$\sigma_{12}^2 = \frac{N_1\sigma_1^2 + N_2\sigma_2^2 + N_1d_1^2 + N_2d_2^2}{N_1 + N_2}$$

$$d_1 = \left|\overline{X}_2 - \overline{X}_{12}\right| = \left|15 - 15.6\right| = -\ 0.6$$

$$d_2 = \left|\overline{X}_2 - \overline{X}_{12}\right| = \left|16 - 15.6\right| = +\ 0.4$$

$$13.44 = \frac{100\,(3)^2 + 150\sigma_2^2 + 100\,(0.6)^2 + 150\,(0.4)^2}{250}$$

$$13.44 = \frac{900 + 150\sigma_2^2 + 36 + 24}{250}$$

$$1540\,\sigma_2^2 = 3{,}360 - 960 = 2{,}400 \Rightarrow \sigma_2^2 = 16 \Rightarrow \sigma_2 = 4$$

The standard deviation of the second sub-group is 4.

Example 47:

A company paid bonus to its employees as under:

Monthly salary (Rs.)	***Bondus paid (Rs.)***	***Monthly Salary (Rs.)***	***Bonus paid (Rs.)***
100-120	*500*	*180-200*	*900*
120-140	*600*	*200-220*	*1000*
140-160	*700*	*220 and over*	*1100*
160-180	*800*		

The actual salaries of the employees were as given below:

Re. : 205, 190, 195, 218, 187, 168, 250, 168, 190, 168, 170, 175, 178, 175, 150, 125, 148, 168, 156, 145, 125, 110, 162, 130, 150, 184.

Restate the data in the form of a frequency distribution and find out:

(i) *The total bonus paid.*

(ii) *The average salary paid per empoyee.*

(iii) *The standard deviation of the distribution.*

Solution:

For finding out the total bonus paid first we will clasify the data and the data and then determine the desired amount.

Monthly Salary (Rs.)	*Tally Bars*	*f*	*Bonus X*	*fx*
100-120	|	1	500	500
120-140	|||	3	600	1800
140-160	|||	5	700	3500
160-180	|||| ||||	9	800	7200
180-200	||||	5	900	4500
200-220	||	3	1000	2000
220 and over	|	1	1100	1100
		Total = 26		åfX = 20600

(i) Total Bonus paid = Rs. 20,600

Calculation of Average Salary Paid and the Standard Deviation

Salarly	**m**	**(m – 170)/20** **d**	**f**	**fd**	**fd²**
100-120	110	–3	1	–3	9
120-140	130	–2	3	–6	12
140-160	150	–1	5	–5	5
160-180	170	0	9	0	0
180-200	190	+1	5	+5	5
200-220	210	+2	2	+4	8
220-240	230	+3	1	+3	9
			N = 26	$\sum$fd = – 2	$\sum$fd² = 48

$$\overline{X} = A + \frac{\sum fd}{N} \times i$$

A = 170, $\sum$fd = – 2, N = 26, i = 20

$$\overline{X} = 170 - \frac{2}{26} \times 20 = 170 - 1.536 - 1.538 = 168.462 \text{ or } 168.46$$

$$\sigma = \sqrt{\frac{\sum fd^2}{N} - \left(\frac{\sum fd}{N}\right)^2} \times i = \sqrt{\frac{48}{26} - \left(\frac{-2}{26}\right)^2} \times 20$$

$$= \sqrt{1.846 - 0059} \times 20 = 1.3565 \times 20 = 27.13$$

Example 48:

The following table gives the marks obtained by a group of 80 students an examination. Calculate the variance.

Marks obtained	*No. of students*	*Marks obtained*	*No. of students*
10 – 14	*2*	*34 – 38*	*10*
14 – 18	*4*	*38 – 42*	*8*
18 – 22	*4*	*42 – 46*	*4*
22 – 26	*8*	*46 – 50*	*6*
26 – 30	*12*	*50 – 54*	*2*
30 – 34	*16*	*54 – 58*	*4*

Solution:

Calculation of Variance

Marks	*m.p.*		*(m – 32)/4*		
	m	*f*	*d*	*fd*	fd^2
10 – 14	12	2	–5	–10	50
14 – 18	16	4	–4	–16	64
18 – 22	20	4	–3	–12	36
22 – 26	24	8	–2	–16	32
26 – 30	28	12	–1	–12	12
30 – 34	32	16	0	0	0
34 – 38	36	10	+1	+10	10
38 – 42	40	8	+2	+16	32
42 – 46	44	4	+3	+12	36
36 – 50	48	6	+4	+24	96
50 – 54	52	2	+5	+10	50
54 – 58	56	4	+6	+24	144
		N = 80		Σ fd = 30	$\Sigma fd^2 = 562$

For details please refer to chapter on 'Theoretical Distributions'.

$$\text{Variance} = \left[\frac{\Sigma fd^2}{N} \left(\frac{\Sigma fd}{N}\right)^2\right] \times i$$

$\Sigma fd^2 = 562$, $\Sigma fd = 30$,

$N = 80$, $i = 4$

Substituting the values

$$\text{Variance} = \left\{\frac{562}{80} - \left(\frac{30}{80}\right)^2\right\} \times 4^2 = (7.025 - 0.141) \times 16$$

$$= \mathbf{6.884 \times 16 = 110.44.}$$

Example 49:

In the following table is given the number of companies belonging to two areas A and B according to the amount of profits eanrned by them. Draw in the same diagram their Lorenz curves and interpet them.

Profits earned Rs. '000	***No. of Companies***	
	Area A	***Area B***
6	*6*	*2*
25	*11*	*38*
60	*13*	*52*
84	*14*	*28*
105	*15*	*38*
150	*17*	*26*
170	*10*	*12*
400	*14*	*4*

Solution:

Calculations for Drawing the Lorenz Curve

Rs. '000	*Cumulative Profit*	*Cumulative Percentage*	*No. of Companies*	*Cumulative Number*	*Cumulative Percentage*	*No. of Companies*	*Cumulative Number*	*Cumulative Per-*
6	6	0.6	6	6	6	2	2	1
25	31	3.1	11	17	17	38	40	20
60	91	9.1	13	30	30	52	92	46
84	175	17.5	14	44	44	28	120	60
105	280	28.0	15	59	59	38	158	79
150	430	43.0	17	76	76	26	184	92
170	600	60.0	10	86	86	12	196	93
400	1000	100.0	14	100	100	4	200	100

Example 50:

Prepare a frequency table with each class interval of 10 kg. and first class interval as 40-50. Also find out the coefficient of variation:

72	74	40	60	82	115	41	61	65	83
53	110	46	84	50	67	78	79	56	65
68	69	104	80	79	79	52	73	59	81
66	49	77	90	84	76	42	64	64	70
72	50	79	52	103	96	51	86	78	94

Solution:

Class groups	*Tally*	*Frequency f*	*m.p. m*	*(m – 75)/10 d*	*fd*	*fd²*
40-50		4	45	–3	–12	36
50-60		8	55	–2	–16	32
60-70		9	65	–1	–9	9
70-80		16	75	0	0	0
80-90		6	85	+1	+6	6
90-100		3	95	+2	+6	12
100-110		2	105	+3	+6	18
1100-120		2	115	+4	+8	32
		N = 50			Σfd = – 11	Σfd² = 145

$$\text{C.V.} = \frac{\sigma}{X} \times 100$$

$$\overline{X} = A + \frac{\sum fd}{N} \times i = 75 - \frac{11}{50} \times 10 = 75 - 2.2 = 72.8$$

$$\sigma = \sqrt{\frac{\sum fd^2}{N} - \left(\frac{\sum fd}{N}\right)^2} \times i = \sqrt{\frac{145}{50} - \left(\frac{11}{50}\right)^2} \times 10$$

$$= \sqrt{2.9 - 0.0484} \times 10 = 1.6887 \times 10 = 16.887$$

$$\text{C.V} = \frac{16.887}{72.8} \times 100 = 23.2 \text{ percent.}$$

Example 51:

Mean and standard deviation of the foolowing continuous series are 31 and 15.9 respectively. The distribution after taking step deviations is as follows:

d	*−3*	*−2*	*−1*	*0*	*1*	*2*	*3*
f	*10*	*15*	*25*	*25*	*10*	*10*	*5*

Determine the actual class intervals.

Solution:

In order to ascertain the calss groups we need two values-the values-the class interval and the assumed mean. From the formula for finding out standard deviation we can determine the class interval and form the formula for calculating mean we can determine the assumed mean.

Computation for Determining Class Groups

d	f	fd	fd²
−3	10	−30	90
−2	15	−30	60
−1	25	−25	25
0	25	0	0
+1	10	+10	10
+2	10	+20	40
+3	5	+15	45
	N = 100	Σfd = − 40	Σfd² = 270

$$\sigma = \sqrt{\frac{\sum fd^2}{N} - \left(\frac{\sum fd}{N}\right)^2} \times i$$

$\Sigma fd^2 = 270$, $\Sigma fd = -40$, $N = 100$ and $\sigma = 15.9$

$$15.9 = \sqrt{\frac{270}{100} - \left(\frac{-40}{100}\right)^2} \times i$$

$$\Rightarrow \quad \sqrt{2.7 - 0.16} \times i = 1.59 \times i$$

$$i = \frac{15.9}{1.59} = 10$$

$$\overline{X} = A + \frac{\sum fd}{N} \times i$$

$$\overline{X} = 31, \ \Sigma fd = -40, \ i = 10$$

$$31 = A - \frac{40}{100} \times 10$$

$\Rightarrow$ $A - A = 31$

$\Rightarrow$ $A = 31 + 4 = 35.$

Hence assumed mean from which deviations have been taken =- 35 and class intervals is 10. The lower and upper limits of this class would be 30 and 40. This class will correspond to zero (0) in the question given. The class preceding to this would be 20-30 and the class succeeding to this 40-50 and likewise we get other classes. Thus actual class group will be as follows:

Class group	*Frequency*	*Class group*	*Frequency*
0-10	10	40-50	10
10-20	15	50-60	10
20-30	25	60-70	5
30-40	25		

Example 52

The mean and standard deviation of normal distribution are 60 and 5 respecitvely. Find the inter-quartile range and the mean deviation of the distribution:

Solution:

Given $\overline{X} = 60, \ \sigma = 5.$

We have to find out the inter-quartile range and the mean deviation.

$$\text{M.D.} = \frac{4}{5}, \sigma = \frac{4}{5} \times 5 = 4$$

$$\text{Q.D.} = \frac{2}{3}, \sigma = \frac{8}{3} \times 5 = \frac{10}{3}$$

$$\frac{Q_3 - Q_1}{2} = \frac{10}{3} \Rightarrow Q_3 - Q_1 = \frac{20}{3} = 6.67$$

Hence inter-quartile range = 6.67.

Example 40:

Praticulars regarding the income of two villages are given below:

	Village X	*Village Y*
Number of people	*600*	*500*
Average income (in Rs.)	*175*	*186*
Variance of income (in Rs.)	*100*	*81*

(i) *In which village is the variation in income greater?*

(ii) *What is the combined standard deviation of the village X and village Y put together?*

Solution:

(i) For finding out the village in which the variation in income is 'greater we have to compare the coefficent of variation.

Village X	***Village Y***
$C.V. = \frac{\sigma}{X} \times 100$	$C.V. = \frac{\sigma}{X} \times 100$
$\frac{10}{175} \times 100 = 5.714$	$\frac{9}{186} \times 100 = 4.839$

Variation in income is greater in village X

$$\sigma_{12} = \sqrt{\frac{N_1\sigma_1^2 + N_2\sigma_2^2 + N_1d_1^2 + N_2d_2^2}{N_1 + N_2}}$$

$$d_1 = (\overline{X}_1 - \overline{X}_{12}),\ d_2 = (\overline{X}_2 - \overline{X}_{12})$$

$$\overline{X}_{12} = \frac{N_1\overline{X}_1 + N_2\overline{X}_2}{N_1 + N_2} = \frac{(600 \times 175) + (500 \times 186)}{600 + 500} = \frac{1{,}05{,}000 + 93{,}000}{1100}$$

$$= 180$$

$$d_1 = |175 - 180| = 5,\ d_2 = |186 - 180| = 6$$

$$\sigma_{12} = \sqrt{\frac{600\,(100) + 500\,(81) + 600\,(5)^2 + 500\,(6)^2}{1100}}$$

$$= \sqrt{\frac{66{,}000 + 40{,}500 + 15{,}000 + 18{,}000}{1100}} = 11.01.$$

Example 53:

You are incharge of rationig in a State affected by food shortage. The following reports arrive from your lecal investigators:

Daily calorie value of food available per adult during current period:

Area	***Mean***	***Standard Deviation***
X	*2,500*	*500*
Y	*2,200*	*300*

The estimated requirement of an adult is taken at 3,000 calories daily and the absolute minimum at 1250. Comment on the reported figre and determine which in your opinion needs more urgent attention.

Solution:

In a polulation $\overline{X} + 3\sigma$ covers 99.73%, *i.e.*, almot all cases. The limits on the basis of the information given to us should be:

Area X: $\overline{X} \pm 3\sigma = = 2{,}500 \pm 3 \times 500 = 1{,}000$ to 4,000

Area Y: $\overline{X} \pm 3\sigma = = 2{,}200 \pm 3 \times 300 = 1{,}300$ to 3,100

It is clear from these limits that in area × there are some people who do not get even 1,250 calories which are regarded as bare minimum whereas in area Y every body is getting above the minimium. Hence area X needsmore urgent attention.

Example 54:

The arithmetic mean and standard deviation of a series of 20 items were calculated by a student as 20 cm and 5 cm respectively. But while calculating an item 13 was misread as 30. Find the correct arithmetic meen and standard deviation.

Solution:

Clculation of Correct Mean

$$\overline{X} = \frac{\sum X}{N} \Rightarrow N\overline{X} = \sum X \; N = 20, \; \overline{X} = 20$$

$$\Rightarrow \sum X = 20 \times 20 = 400 \text{ Correct } \sum X = 400 - 30 + 13 = 383$$

$$\text{Correct } \overline{X} = \frac{383}{20} = 19.15$$

Calculation of Correct Standard Deviation

$$\sigma^2 = \frac{\sum X^2}{N} - (\overline{X})^2 \Rightarrow \sigma^2 = \frac{\sum X^2}{N} - (20)^2$$

$$20 \times 20 = \sum X^2 - 400 \times 20 \quad \Rightarrow \quad \sum X^2 = 8500$$

$$\text{Correct } \sum X^2 = 8500 - (30)^2 + (13)^2 = 8500 - 900 + 169 = 7769.$$

$$\text{Correct } \sigma^2 = \sqrt{\frac{\text{Correct} \sum X^2}{N} - (\text{Correct } \overline{X})^2}$$

$$= \sqrt{\frac{7769}{20} - (19.15)^2} = \sqrt{383.45 - 366.72} = \sqrt{21.73} = 4.66.$$

Thus the correct mean is 19.15 and correct standard deviation 4.66.

Example 55:

A collar manufcturer is considering the production of a new style of collar to attract young men. The following statisics of neck circumferences are available based upon measurements of a typical group of college students:

Mid-value (in inches)	*12.0*	*12.5*	*13.0*	*13.5*	*14.0*	*14.5*	*15.0*	*15.5*	*16.0*
No. of students	*2*	*16*	*36*	*60*	*76*	*37*	*18*	*3*	*2*

Compute the standard deviation and use the criterion $\overline{X} + 2s$ *where s is the standard deviation and* $\overline{X}$ *is the arithmetic mean to determine the largest and smallest size of the collar he should make in order to meet the needs of practically all the customers bearing in mind that collars are worn on an average 1 inch loger tan neck size.*

Solution:

Calculation of Mean and Standard Deviation

Mid-value in Inches	No. of Staudents f	(m – 14)/0.5 d	fd	fd²
12.0	2	–4	–8	32
12.5	16	–3	–48	144
13.0	36	–2	–72	144
13.5	60	–1	–60	60
14.0	76	0	0	0
14.5	37	+1	+37	37
15.0	18	+2	+36	72
15.5	3	+3	+9	27
16.0	2	+4	+8	32
	N = 250		Σfd = – 98	Σfd² = 548

$$\overline{X} = A + \frac{\sum fd}{N} \times i = 14 - \frac{98}{250} \times 0.5 = 13.8$$

$$\sigma = \sqrt{\frac{\sum fd^2}{N} - \left(\frac{\sum fd}{N}\right)^2} \times i = \sqrt{\frac{548}{250} - \left(\frac{-98}{250}\right)^2} \times 0.5$$

$$= \sqrt{2.192 - 0.154} \times (5) = \sqrt{2.038} \times 0.5 = 1.43 \times 0.5 = 0.715$$

Largest neck size = $\overline{X} + 3\sigma = 13.8 + 3(0.715) = 13.8 + 2.145 = 15.945.$

Smallets neck size = $\overline{X} - 3\sigma = 13.8 - (0.715) = 13.8 - 2.145 = 11.655.$

Since collars are worn on an average 1 inch longer than the neck size, we shold add 0.5 to these limits. Thus the smallest and largest size of collar should be

$$(11.655 + 0.5) \text{ and } (15.945 + 0.5) = 12.155 \text{ and } 16.445.$$

Thus the smallest size of collar should be 12.2 inches long and largest 16.4 inches long.

Example 56:

Find the missing information from the following:

	Group I	*Group II*	*Group III*	*Combined*
Number	*50*	*?*	*90*	*200*
Standard Deviation	*6*	*7*	*?*	*7.746*
Mean	*113*	*?*	*115*	*116*

Solution:

Finding number of Observations in the Second Group.

Let N_1, N_2 N_5 denote number of observations in the 1st, 2nd, 3rd group respectively.

We are given $N_1 + N_2 + N_3 = 200$

$$N_1 = 50, N_3 = 90 \quad \therefore \quad N_1 + N_2 = 140$$

$$\therefore \qquad N_2 = 200 - 140 = 60.$$

Finding Mean of Second Group

Let $\overline{X}_1$, $\overline{X}_2$, $\overline{X}_3$, denote mean of first, second and third group respectively.

$$\overline{X}_{123} = \frac{N_1\overline{X}_1 + N_2\overline{X}_2 + N_3\overline{X}_3}{N_1 + N_2 + N_3}$$

$$\overline{X}_{123} = 116, N_1 + N_2 + N_3 = 200, \ \overline{X}_1 = 113, \overline{X}_2 = 115$$

We have to find $\overline{X}_2$

Substituting the given values

$$116 = \frac{(50)\,113 + 60(\overline{X})_2 + 90\,(115)}{200}$$

$$116 \times 200 = 5650 + 60\,\overline{X}_2 + 10350$$

$$60\,\overline{X}_2 = 23{,}200 - 16{,}000 = 7200 \therefore \quad \overline{X}_2 = \frac{7{,}200}{60} = 120$$

Finding Standard Deviation of 3rd Group

$$\sigma_{123} = \sqrt{\frac{N_1\sigma_1^{\,2} + N_2\sigma_2^{\,2} + N_3\sigma_3^{\,2} + N_1 d_1^{\,2} + N_2 d_2^{\,2} + N_3 d_3^{\,2}}{N_1 + N_2 + N_3}}$$

$$\sigma_{123} = 7.746,\ N_1 = 50,\ \sigma_1 = 6,\ N_2 = 60,\ \sigma_3 = 7,\ N_2 = 90$$

$$d_1 = \left|\overline{X}_1 - \overline{X}_{123}\right| = (113 - 116) = 3$$

$$d_2 = \left|\overline{X}_3 - \overline{X}_{123}\right| = (120 - 116) = 4$$

$$d_3 = \left|\overline{X}_3 - \overline{X}_{123}\right| = (115 - 116) = 1$$

Substituting the values

$$7.746 = \sqrt{\frac{50(6) + 60(7)^2 + 90\sigma_2^{\,2} + 50(3)^2 + 60(4)^2 + 90(1)^2}{50 + 60 + 90}}$$

$$= \sqrt{\frac{1{,}800 + 2{,}940 + 90\sigma_2^{\,2} + 450 + 960 + 90}{2000}} = \sqrt{\frac{6{,}240 + 90\sigma_2^{\,2}}{200}}$$

$$\text{Squaring } (7.746)^2 = \frac{6.240 + 90\sigma_2^{\,2}}{200}$$

$$12,\ 000 = 6{,}240 + 90\sigma_2^{\,2} \Rightarrow \quad 90\sigma_2^{\,2} = 12{,}000 - 6{,}240$$

$$\therefore\ \sigma_2^{\,2} = \frac{5{,}760}{90} = 64 \Rightarrow \sigma_2 = \sqrt{64} = 8$$

Thus the missing values are

$$N_3 = 60,\ \overline{X}_2 = 120,\ \sigma_2 = 8.$$

Example 57:

Find the standard deviation, and coefficient of variation from the following data:

Wages	*No. of workers*	*Wages*	*No. of workers*
Up to Rs. 10	*12*	*Up to Rs. 50*	*157*
" " " 20	*30*	*" " " 60*	*202*
" " " 30	*65*	*" " " 70*	*222*
" " " 40	*107*	*" " " 80*	*230*

Solution:

Calculation of Coefficient of Variation

Wages (Rs.)	*m.p.* *m*	*No of workers*	*(m – 35)/10* *d*	*fd*	*fd²*
0-10	5	12	–3	–36	108
10-20	15	18	–2	–36	72
20-30	25	35	–1	–35	35
30-40	35	42	0	0	0
40-50	45	50	+1	+50	50
50-60	55	45	+2	+90	180
60-70	65	20	–3	+60	180
70-80	75	8	+4	–32	128
		N = 230		Σfd = + 125	Σfd² = 753

$$C.V. = \frac{\sigma}{X} \times 100$$

$$\sigma = \sqrt{\frac{\sum fd^2}{N} - \left(\frac{\sum fd}{N}\right)^2} \times i = \sqrt{\frac{753}{230} - \left(\frac{125}{230}\right)^2} \times 10$$

$$= \sqrt{3.274 + 0.295} \times 10 \sqrt{2.979} \times 10 = 1.726 \times 10 = 17.26$$

$$\overline{X} = A + \frac{\sum fd}{N} \times i = 35 + \frac{125}{230} \times 10 = 35 + 5.43 = 40.\ 43$$

$$C.V. = \frac{17.26}{40.43} \times 100 = 42.69 \text{ percent.}$$

Example 58:

For a group containing 100 observations, the arithmetic mean and standard deviation are 8 and $\sqrt{10.5}$ *. For 50 observations selected from these 100 observations the mean and the standard deviation are 10 and 2 respectively. Find the arithmetic mean and the standard deviation of the other half.*

Solution:

$$N_1 + N_2 = 100, \quad \overline{X}_{12} = 8 \;\; \sigma_{12} = \sqrt{10.5}$$

$$N_1 = 50, \quad \overline{X}_1 = 10, \sigma_1 = 8.$$

We have to find the mean and standard deviation of the other helf.

$$\overline{X}_{12} = \frac{N_1\overline{X}_1 + N_2\overline{X}_2}{N_1 + N_2}; \; 8 = \frac{50 \times 10 + 50\overline{X}_2}{100}$$

$$800 = 500 + 50\,\overline{X}_2 \;\Rightarrow\; 50\,\overline{X}_2 = 300 \;\Rightarrow\; \overline{X}_2 = 6$$

$$\sigma_2{}^2 = \frac{N_1\sigma_1{}^2 + N_2\sigma_2{}^2 + N_1 d_1{}^2 + N_2 d_2{}^2}{N_1 + N_2}$$

$$d_1 = \left|\overline{X}_1 - \overline{X}_{12}\right| = 10 - 8 = 2$$

$$d_2 = \left|\overline{X}_2 - \overline{X}_{12}\right| = 6 - 8 = -2$$

$$10.5 = \frac{50(2)^2 + 50\sigma_2{}^2 + 50(2)^2 + 50(-2)^2}{100}$$

$$10.5 \times 100 = 200 + 50\sigma_2{}^2 + 200 + 200$$

$$50\sigma_2{}^2 = 1050 - 600 = 450$$

$$\sigma_2{}^2 = \frac{450}{50} = 9 \quad \Rightarrow \quad \sigma_2 = 3.$$

Hence the mean of the remaining 50 items is 8 and the standard deviation 3.

Example 59:

In a small town, a survey was conducted in respect of profits made by retail shops. The following resuits were obtained:

Profite or Loss in '000 Rs.	***No. of shops***	***Profit or Loss in '000 Rs.***	***No. of shops***
–4 to –3	*4*	*1 to 2*	*56*
–3 to –2	*10*	*2 to 3*	*40*
–2 to –1	*22*	*3 to 4*	*24*
–1 to – 0	*28*	*4 to 5*	*18*
–0 to –1	*38*	*5 to 6*	*10*

Calculate

(i) The average profit made by a retail shop.

(ii) Total profit by all shops.

(iii) The coefficient of variation of earnings.

Solution:

Calculation of Mean and Standard Deviation

Wages (Rs.) Loss in '000 rs.	*m.p. m*	*No of Shops f*	*(m – 1.5) d*	*fd*	*fd²*
–4 to –3	–3.5	4	–5	–20	100
–3 to –2	–2.5	10	–4	–40	160
–2 to –1	–1.5	22	–3	–66	198
–1 to –0	–0.5	28	–2	–56	112
0 to 1	+0.5	38	–1	–38	38
1 to 2	+1.5	56	0	0	0
2 to 3	+2.5	40	+1	+40	40
3 to 4	+3.5	24	+2	+48	96
4 to 5	+4.5	18	+3	–54	162
5 to 6	+5.5	10	+4	+40	160
		N = 250		Σfd = – 38	Σfd² = 1,066

$$\overline{X} = A + \frac{\sum fd}{A} = 1.5 - \frac{38}{250} = 1.5 = 1348 \Rightarrow \text{Rs. } 1.348$$

$$\sigma = \sqrt{\frac{\sum fd^2}{A} - \left(\frac{\sum fd}{A}\right)^2} = \sqrt{\frac{1066}{250} - \left(\frac{-38}{250}\right)^2}$$

$$= \sqrt{4.2824 - 0.023} = 2.059 \Rightarrow \text{Rs. } 2,059$$

$$\text{C.V} = \frac{\sigma}{X} \times 100 = \frac{2059}{1348} \times 100 = 152.74\%$$

Total profit, *i.e.*, $\Sigma X = N\overline{X} = 250 \times 1.318 = \text{Rs. } 3.37,00$

Example 60:

The mean and standared deviation of 200 items are found to be 60 and 20 respectively. If at the time of calculations, two items were wrongly taken as 3 and 67 instead of 13 and 17, find the correct mean and standard deviation. What is the correct coefficient of variation?

Solution:

We are given $\overline{X} = 60$, $\sigma = 20$, $N = 200$

$$\overline{X} = \frac{\sum X}{N} \Rightarrow 60 = \frac{\sum X}{200} \Rightarrow \sum X = 12{,}000$$

But correct $\sum X = \sum X$ – wrong items + correct items

$$= 12000 - 3 - 67 + 17 = 11960$$

$$\therefore \text{Correct Mean} = \frac{\text{Correct} \sum X}{N} = \frac{11.960}{200} = 59.8$$

Correct Standard Deviation

$$\sigma = \sqrt{\frac{\sum X^2}{N} - (\overline{X})^2} \Rightarrow 10 = \sqrt{\frac{\sum X^2}{200} - (60)^2}$$

$$\text{Squaring } 400 = \frac{\sum X^2}{N} - 3{,}600$$

$$80{,}000 = \sum X^2 = 3600 \times 200$$

$$\sum X^2 = 80{,}000 + 7{,}20{,}000 = 8{,}00{,}000$$

Correct $\sum X^2$ = Incorrect $\sum X^2$ – Wrong Items square + Correct items squares

$$\text{Correct } \sum X^2 = 8{,}00{,}000 - (3)^2 - (67)^2 + (13)^2 + (17)^2$$

$$= 8{,}00{,}000 - 9 - 4{,}489 + 169 + 289 = 7{,}95{,}960$$

$$\text{Correct } \sigma = \sqrt{\text{Correct} \frac{\sum X^2}{N} - (\text{Correct } \overline{X})^2} = \sqrt{\frac{795960}{200} - (59.8)^2}$$

$$= \sqrt{3979.8 - 3576.04} = \sqrt{403.76} = 20.094.$$

Example 61:

A sample of 35 values has mean 80 and standard deviation 4. A second sample of 65 values has mean 70 and standard deviation 5. Find the standard deviation of the combined sample of 100 values.

Solution:

Combined Standard Deviation of two series is given by the formula

$$\sigma_{12} = \sqrt{\frac{N_1\sigma_1^2 + N_2\sigma_2^2 + N_1 d_1^2 + N_2\sigma_2^2}{N_1 + N_2}}$$

$N_1 = 35$, $\overline{X}_1 = 80$, $\sigma_1 - 4$, $\sigma_2 = 5$, $N_2 = 65$, $\overline{X}_2 = 70$, $\sigma_2 = 5$

$$\overline{X}_{12} = \frac{N_1\overline{X}_2 + N_2\overline{X}_2}{N_1 + N_2}$$

$$= \frac{(35 \times 80) + (65 \times 70)}{35 + 65} = \frac{2800 + 4550}{100} = \frac{7350}{100} = 73.5$$

$$d_1 = |\overline{X}_1 - \overline{X}_{12}| = |80 - 72.5| = 6.5$$

$$d_2 = |\overline{X}_2 - \overline{X}_{12}| = |70 - 73.5| = 3.5$$

$$\sigma_{12} = \sqrt{\frac{35(4)^2 + 65(5)^2 + 35(6.5)^2 + 65(3.5)^2}{100}}$$

$$= \sqrt{\frac{560 + 1625 + 1478.75 + 796.25}{100}} = \sqrt{\frac{4460}{100}} = 6.68$$

Example 62:

Calculate deviation and its coefficient of the following distribution of 'collar' measurements:

Mid-value (inches)	*12.5*	*13.0*	*13.5*	*14.0*	*14.5*	*15.0*	*15.5*	*16.0*	*16.5*
No. of students	*4*	*19*	*30*	*63*	*66*	*29*	*18*	*1*	*1*

Solution:

Since we are given the midvalues we first determine the lower and upper limits. Since the difference between first and second midvalue is 0.5, deduct half of it, ie., 25 from first value *i.e.*, lower limit in 12.25 and add 25 to given first value, *i.e.*, it becomes 12.75 etc.

Calculation of Q.D. and its Coefficient

Lass limits	***m.p.*** *m*	*f*	*c.f*
12.25–12.75	12.5	4	4
12.75–13.25	13.0	19	23
13.25–13.75	13.5	30	53
13.75–14.25	14.0	63	116
14.25–14.75	14.5	66	182
14.75–15.25	15.0	29	211
15.25–15.25	15.5	18	119
15.75–16.25	16.0	1	130
16.25–16.75	16.5	1	231

$$Q.D. = \frac{Q_3 - Q_1}{2}$$

$$Q_1 = \text{Size of } \frac{N}{4}\text{th item} = \frac{231}{4} = 57.75\text{th item}$$

Q_1 lies in the calss 13.75 – 14.25

$$Q_1 = L + \frac{N/4 - c.t.}{f} \times i$$

L = 13.75, N/4 = 57.75, c.f. = 53, f = 63, i = 05

$$Q_1 = 13.75 + \frac{57.75 - 53}{63} \times 0.5 = 13.75 + 0.04 = 13.79$$

$$Q_3 = \text{Size of } \frac{3N}{4}\text{th item} = \frac{3 \times 231}{4} = 173.25\text{th item}$$

Q_3 lies in the class 14.25 – 14.75

$$Q_3 = L + \frac{3N/4 - c.f.}{f} \times i$$

L = 14.25, 3N/4 = 173.25, c.f. = 116, f = 66, i = 0.5

$$Q_3 = 14.25 + \frac{173.25 - 116}{66} \times 0.5$$

$$= 14.25 + 0.433 = 14.683$$

$$Q.D = \frac{Q_3 - Q_1}{2} = \frac{14.683 - 13.79}{2} = 0.447$$

$$\text{Coeff. of Q.D.} = \frac{Q_3 - Q_1}{Q_3 + Q_1}$$

$$= \frac{14.683 - 13.79}{14.683 + 13.79} = \frac{0.893}{28.473} = 0.031$$

Example 63:

A consignment of 180 articies is classified according to the size of the article as under. Find the standard deviation and its coefficient:

Measurement	*No. of articles*	*Measurement*	*No. of articles*
More than 80	*5*	*More than 30*	*150*
More than 70	*14*	*More than 20*	*170*
More than 60	*34*	*More than 10*	*176*
More than 50	*65*	*More than 0*	*180*
More than 40	*110*	*More than 90*	*0*

Solution:

This is a cumulative frequency distribution. First convert it to a simple frequency distribution in an ascending order.

Measurement	f	m.p. m	(m – 45)/10 d	fd	fd²
0-10	4	5	–4	–16	64
10-20	6	15	–3	–18	54
20-30	20	25	–2	–40	80
30-40	40	35	–1	–40	40
40-50	45	45	0	0	0
50-60	31	55	+1	+31	31
60-70	20	65	+2	+40	80
70-80	9	75	+3	+27	81
80-90	5	85	+4	+20	80
	N = 180			Σfd = 4	Σfd² = 510

$$\sigma = \sqrt{\frac{\sum fd^2}{N} - \left(\frac{\sum fd}{N}\right)^2} \times i = \sqrt{\frac{510}{180} = \left(\frac{4}{180}\right)^2} \times 10$$

$$= \sqrt{2.833 - 0.0005 \times 10} = 1.683 \times 10 = 16.83$$

Coeff. of standard deviation. It can be obtained by dividing standard deviation by mean.

$$\overline{X} = A + \frac{\sum fd}{N} \times i = 45 + \frac{4}{180} \times 10 = 45 + 222 = 45.222$$

$$\text{Coeff. of S.D.} = \frac{\sigma}{X} = \frac{16.83}{45.222} = 0.372$$

EXERCISES

1. Find the coefficient of variation from the following:

   (a) $\Sigma X = 250$, $N = 10$ $s = 8$

   (b) Mean = 40, Variance = 25.

   (c) $\Sigma dx^2 = 272$, $\Sigma dx = 25$, $N = 100$.

   Assumed mean = 4.

2. An analysis of the monthly wages paid to workers in two firms A and B, belonging to the same industry, gave the following results:

		Firm A	*Firm B*
No. of workers	:	160	150
Average wage	:	560	575
Variance of wage distribution	:	400	625

Find out:

(a) Which firing pays larger amount as monthly wages?

(b) In which firm is there greater variability in individual in individual wages?

(a) Total Wage bill (A) = 89,600, (B) = 86.250. [(A)]

(b) C.V. (A) = 357, C.V. (B) = 4.35 [(B)]

3. (a) The index number of prices of Cotton and Coal shares in 1988 were as under:

Month	*Jan.*	*Feb.*	*March*	*April*	*May*	*June*	*July*	*Aug.*	*Sept.*	*Oct.*
Cotton	188	178	173	164	172	184	184	185	211	217
Coal	131	130	130	129	129	120	127	127	130	137

Which of the two shares you consider more variable in price?

(b) The variable x takes only two values x_1 and x_2 with frequencies f_1 and f_2 respectively. If S be the standard deviation of p, show that:

$$S^2 = f_1 f_2 \left(\frac{x_1 - x_2}{f_1 + f_2} \right)^2$$

4. Following are the data for marks obtained by students in a paper. The top 20% students will qualify for a prize. What is the lower limit of marks, above which the student will get the prize?

Marks	*No. of students*	*Marks*	*No. of students*
0–10	5	50–60	10
10–20	7	60–70	4
20–30	8	70–80	4
30–40	10	80–90	2
40–50	10		

5. (a) The following table gives the height of students in a class. Find out the quartile deviation.

Height (inches) :	50–53	53–56	56–59	59–62	62–65	65–68
No. of students :	2	7	24	27	13	3

(b) Calculate the standard deviation for the following data:

X:	50	60	70	80	90	100	110	120
f:	14	40	54	46	26	12	6	2

6. A student obtained the mean and standard deviation of 100 observations as 40 and 5.1 respectability. It was later found that one observation was wrongly coupled as 50, the correct figure being 40. Find the correct mean and standard deviation.

7. Tick the correct answer:

   I. Coefficient of quartile deviation is calculated by the formula:

   (i) $\frac{Q_2 + Q_1}{4}$, (ii) $\frac{Q_3 + Q_1}{2}$, (iii) $\frac{Q_3 - Q_1}{Q_3 + Q_1}$, (iv) $\frac{Q_2 +}{Q_3 -}$

   (v) None of these

   II. Mean ± 3σ covers: (i) 93.37% items, (ii) 90%, (iii) 99.73%, (v) all the items.

   III. Quartile deviation is : (i) 4/5 σ, (ii) 3/2 σ, (iii) 2/3 σ, (iv) 4/5 σ, (v) None of these.

   IV. Coefficient of variation is calculated by the formula:

   (i) $\frac{\overline{X}}{s} \times 100$ (ii) $\frac{\overline{X}}{s}$, (iii) $\frac{s}{\overline{X}} \times 100$,(iv) $\frac{s}{\overline{X}} \times 10$(

   (v) None of these.

   V. The measure of variation that is least affected by extreme observations is: (i) Range, (ii) Mean deviation, (iii) Standard deviation, (iv) Quartile deviation, (v) All of these,

   [I. (iii), II (v), III, (i), IV, (iv), V, (ii)]

8. Find the actual class group from the information given below:

d :	–3	–2	–1	0	1	2	3	4
f :	15	15	2	22	25	10	5	10

You are told $\overline{X}$ of the distribution is 36.16 and σ = 19.76.

[A = 35, i = 10: Class groups 0–10, 10–20, ... 70–80]

9. Define dispersion as a concept. State the measures in use for it. Calculate one of these fr the following data:

Infection

Score (%) :	0–10	10–20	20–30	30–40	40–50
No. of affected fields :	2	3	8	15	26
Infection score (%) :	50–60	60–70	70–80	80–90	90–100
No. of affected fields :	30	6	5	4	1

[σ = 16.9]

10. A survey of taxi drivers in Mumbai gave the following information:

Kilometres driven per month	*Number of drivers*	*Kilometres driven per month*	*Number of drivers*
2,000–2,999	5	6,000–6,999	16
3,000–3,999	12	7,000–7,999	9
4,000–4,999	18	8,000–8,999	3
5,000–5,999	38		

Find (i) the range which contains the middle 50% of the drivers, and (ii) the coefficient of variation.
[(ii) C.V. = 54.19]

11. Find out which of the following batsmen is more consistent in scoring. Would you also accept him as a better run getter? Why?

Batsman 'A' :	5	7	16	27	39	53	56	61	80	101	105
Batsman 'B' :	0	4	16	21	41	43	57	78	83	90	95

(b) Initially, there were 9 workers, all being paid a uniform wage. Later, a 10th worker is added whose wage rate is Rs. 20 less than for the others. Compute:

(i) the effect on the mean wage.

(ii) Standard deviation of wages for the group of 10 workers.

12. The following table shows the monthly expenditure of 60 hostelers of a co-ed, hostel on refreshment:

Expenditure (in Rs.)	*No. of students*	*Expenditure (in Rs.)*	*No. of students*
170–180	4	220–240	4
180–190	6	240–260	2
190–200	7	260–280	8

200–210	12	280–300	7
210–220	10		

Calculate arithmetic mean and standard deviation of the above data and examine the implications of results on expenditure planning of the hostel.

13. The following data pertain to two workers doing the same job in a factory:

	Worker A	***Worker A***
Mean time of completing the job (minutes)	40	42
Standard deviation (minutes)	8	6

Who is the more consistent worker?

14. Prices of a particular item in 10 years in two cities are given below, which city has more stable prices?

City A :	55	5452	53	5658	52	5051	49
City B :	108	107105	105	106107	104	103104	101

15. Calculate the coefficient of variation of the following two series and show which series is more variable:

Weight in kg. :	0–10	10–20	20–30	30–40	40–50	50–60	60–70
Class A: f :	1	2	9	8	5	4	1
Class B: f :	1	3	7	8	7	3	1

16. (a) Following data represent life of two models of refrigerators A and B:

Life (No. of years)	***Refrigerator***
Model A	***Model B***
0–25	2
2–416	7
4–613	12
6–87	19
8–105	9
10–124	1

Find the average life of each model. What model has greater uniformity? Also obtain mode for both models.

(b) For a group of 50 male workers, the mean and standard deviation for wages are Rs. 63 and Rs. 9 respectively. For a group of 40 female workers, these are Rs. 54 and Rs. 6 respectively. Find the combined mean and standard deviation.

17. Following are the records of two players regarding their performance in cricket matches:

Score of Player A:	48	52	55	60	65	45	63	70
Score of Player :	33	35	80	70	100	15	11	25

(a) Which player has scored more on an average?

(b) Which player is more consistent in his performance.

18. Claculate mean deviation from the median for the following data:

Age (years)	4-6	6-8	8-10	10-12	12-14	14-16	16-18
No. of Students:	30	90	120	150	80	60	20

19. (a) The following table gives the number of finished articles turned out daily by the different number of workers in the factory. Find the mean value and the standard deviation of output of finished articles daily.

No. of Articles:	18	19	20	21	22	23	24	25	26	27
No. of Workers:	3	7	11	14	18	17	13	8	5	4

(b) Explain, with a suitable example, the term 'Variation'. Mention some common measures of variation and describe the one which you think to be most important of them.

20. (a) The following are the scores of Manoj and Rajeev for 8 innings.

Manoj :	12	115	76	42	7	19	49	80
Rajeev :	47	12	76	73	24	51	63	54

Who of the two is a nor consistent batsman?

(b) Calculate the mean deviation from mean from the following series and find its cocfficient:

Marks :	0–10	10–20	20–30	30–40	40–50
No. of Students :	5	8	15	16	6

21. (a) The mean and standard deviation of series of seventeen items are 25 and 5 respectively. While calculating these measures, a measurement 53 was wrongly read as 35, Correct this error and find out the correct standard deviation.

(b) The daily temperature recorded in a city in Russia in a year is given below:

Temperature 'C	*No. of days*	*Temperature 'C*	*No. of days*
–40 to –30	10	0 to 10	65
–30 to –20	28	10 to 20	180
–20 to –10	20	20 to 30	10
–10 to 0	42		

Calculate the mean, standard deviation and coefficient of variation.

22. For two firms A and B belonging to same industry, the following details are available

	Firm A	*Firm B*
Number of Employees :	100	200
Average wage per month :	Rs. 240	Rs. 170
Standard deviation	Rs. 6	Rs. 8

Find: (i) Which firm pays out larger amount as monthly wages?

(ii) Find average monthly wage and the standard deviation of the wages of all employees in both the firms.

[(i) B, (ii) B, (iii) $\overline{X}_{12}$ = 193.33]

23. (a) The mean of two samples of size 50 and 100 respectively are 54.1 and 50.3 and standard deviations are 8 and 7. Find the mean and the standard deviation of the sample of size 150 obtained by combining the two samples.

(b) Two samples of size 40 to 50 have the same mean 53, but different standard deviation of the combined sample of size 90.

(c) For a group of 200 candidates, the mean and standard deviations were found the be 40 and 15. Later on, it was discovered that the score 43 was missed as 53. Find the correct mean and standard deviation after correction.

(d) Mean and standard deviation of 200 items are found to be 60 and 20. If at the time of calculations two items are wrongly taken as 3 and 67 instead of 13 and 17, find the correct mean and standard deviation.

[(a) $\overline{X}_{12}$ = 51.57, σ = 7.56: (b) σ_{12} = 14: (c) 39.95: 14.97: 14.97; (d) 59.8, 10.09]

24. Comment briefly on the following statements:
    (a) The median is the point about which the sum of the squared deviations is minimum.
    (b) A computer found that the standard deviation of a set of 40 observations, whose values ranged between 116 and 136, is 22,
    (c) The range is the mean perfect measure of variability because it includes all the measurements.
    (d) After the settlement of dispute, the average weekly wage in a factory had increased from Rs. 8 to 12 and the standard deviation had increased from 1 to 15. After settlement, the wages has become higher and more uniform.
25. Find the mean, median and standard deviation of the weight of bullets in guns as given in the following table:

Variable	*Frequency*	*Variable*	*Frequency*
210–215	8	230–235	14
215–220	13	235–240	10
220–225	16	240–245	7
225–230	29	245–250	3

[$\overline{X}$ = 227.55; Med, = 227.24; σ = 8.73]

26. The profits (in Rs. lakhs) earned by 100 companies during 1997-98 are shown below:

Profits	*No. of Companies*	*Profits*	*No. of Companies*
20–30	4	60–70	15
30–40	8	70–80	10
40–50	18	80–90	8
50–60	29	90–100	7

Compute (a) Mean, (b) Median, (c) Standard deviation.

[$\overline{X}$ = 59.1: Med = 56.67: σ = 17.56]

27. Calculate the mean and standard deviation of the following distribution:

Age (Years) :	25–30	30–35	35–40	40–35	45–50	50–55
No of workers :	70	51	47	31	29	22

28. (a) Your are supplied the following data about height of boys and girls in a college:

	Number	*Average Height*	*Variance*
Boys	72	68"	9"
Girls	38	61"	4"

You are required to find out (i) the combined mean and S.D. of heights of boys and girls taken together, and show (ii) whose height is more variable?

(b) Find mean deviation and its coefficient of the following series:

Size :	10	11	12	13	14
Frequency :	5	12	18	12	3

29. (a) Prove that the some of the squared deviations is least when taken from the mean.

(b) Dispersion is known as the average of the second order. Give reasons.

30. (a) Which measure of dispersion would be most useful for:

(i) a social worker,

(ii) an actuary.

(iii) A public relations man for industry, and

(iv) a spokesman for organised labour?

(b) What is coefficient of dispersion? What purpose does it serve?

31. (a) What is a measure of dispersion? Discuss four important measures of spread indicating their uses.

(b) Prove that the standard deviation is independent of the change of origin but not of scale.

32. (a) Explain the concept of dispersion in statistical analysis. Describe its various measures and discuss their merits and demerits.

(b) Write down the formula for mean deviation and standard deviation in a discrete series.

33. During the 10 weeks of a session, the marks scored by two candidates, Jayanth and Vasanth, taking the computer programs course are given below:

Jayanh :	58	59	60	54	65	66	52	75	69	62
Vasanth :	87	89	78	71	73	84	65	66	56	46

(a) Who the better score. Jayanth or Vasanth?

(b) Who is more consistent?

34. From some financial statistics, it is found that the monthly average of electricity charges was Rs. 2.460 and S.D. Rs. 120. The monthly average of direct wages was Rs. 42,000 and S.D. Rs. 1,200. State, which is more variable?

35. Based on the frequency distribution given below, compute the following statistical measures to characterise the distribution:

    (i) Coefficient of variation (ii) Inter-quartile range, (iii) Modal value.

Annual tax paid (Rs. Thousand)	*No. of Managers*	*Annual tax paid (Rs. Thousand)*	*No. of managers*
5–10	18	25–30	20
10–15	30	30–35	12
15–20	46	35–40	6
20–25	28		

37. (a) What measures of dispersion will you choose for:

    (i) the movement of prices in stock market,

    (ii) reporting on the rainfall in a certain region, and

    (iii) a frequency distribution with an open interval either at the beginning or at the end.

    (b) Correct the following statements:

    (i) the mean deviation of a frequency distribution about an origin is minimum when the origin is the mean.

    (ii) the average deviation of a distribution is greater that standard deviation.

37. Calculate S.D. and C.V. for the following frequency distribution:

Class	*Frequency*	*Class*	*Frequency*
4–8	11	24–28	9
8–12	13	28–32	17
12–16	16	32–36	6
16–20	14	36–40	4
20–24	14		

38. A factory produces two types of electric lamps. A and B. In an experiment relating to their life, the following results were obtained:

Length of life (in hrs.)	*No. of Lamps A*	*No. of Lamps B*
500–700	5	4
700–900	11	30
900–1,100	26	12
1,100–1,300	10	8
1,300–15,00	8	6
Total	60	60

Compare the variability of the two varieties using coefficient of variations.

39. From the following figures, determine the percentage of case which lie outside the mean at distance $\overline{X} \pm 1\sigma$, $\overline{X} \pm 2\sigma$, $\overline{X} \pm 3\sigma$.

115	117	121	125	116	120	118	117	119	116
122	124	123	118	120	118	126	127	122	123

[$\overline{X} \pm \sigma$ = 15%, $\overline{X} \pm 2\sigma$ = 85% and $X \pm 3\sigma$ = nil]

40. Examine the different methods of measuring relationship between two variables. Point out the usefulness of each method taking suitable examples.

41. (a) "Averages and measures of dispersion are useful in understanding a frequency distribution". Educate the statement giving illustrations

    (b) What is a Lorenz Curve? How it is useful in measuring become inequalities between two regions?

42. (a) What are measures of disporting? Examine the qualities of standard deviation as a measure of dispersion.

    (b) Why is standard deviation considered to be the most reliable measure of dispersion?

43. (a) Critically examine the different methods of measuring vacation.

    (b) Describe the relative measures of dispersion.

44. Distinguish between mean deviation and standard deviation Which is considered better and why?

45. (a) What are the properties of a good measure of variation?

    (b) Mention some common measures of dispersion and explain the one which you think to be most important of them.

46. (a) Determine the standard deviation for the following frequency distribution:

Wages per day :	2–4	4–6	6–8	8–10	10–12	12–14	Total
Number of workers :	9	12	18	16	8	7	72

(b) Calculate the quartile deviation for the following frequency distribution:

X :	60	62	64	66	68	70	72
Frequency :	12	16	18	20	15	13	9

47. (a) From the following information, find the standard deviation of x and y variables:

$\Sigma x = 235$, $\Sigma y = 250$, $\Sigma x^2 = 6750$, $\Sigma y^2 = 6840$, $N = 10$

(b) Find the coefficient of variation if variance is 16, number of items is 20 and sum 7 of the items is 160.

(c) From the following data, calculate (i) the coefficient of range, (ii) inter-quartile range, (iii) percentile range.

Marks :	5–9	10–14	15–19	20–24	25–29	30–34	35–39
No. Students:	1	3	8	5	4	2	2

48. Fill in the blanks:

(i) When mean is 79 and variance is 64. C.V.......

(ii) If in a series the coefficient of variation is 20 and mean 40. The standard deviation shall be.......

(iii) If $Q_1 = 30$ and $Q_2 = 50$, the coefficient of quartile deviation shall be.....

(iv) If the coefficient of variation of distribution is 50 and its standard deviation 20, the arithmetic mean shall be.....

(v) Quartile deviation is ...... of standard deviation.

Ans. [(i) 10.26, (ii) 8, (iii) 0.21, (iv) 40, (v) 0.6745]

49. Indicate when the following statements are True or False:

(i) Range is the best measure of dispersion. T/F

(ii) Absolute measure of dispersion can be used for the purpose of comparison. T/F

(iii) Quartile deviation is more suitable in case of open end distributions. T/F

(iv) Standard deviation can be calculated from any average. T/F

(v) There is no difference between variance and coefficient of variation. T/F

(vi) Mean deviation is least when deviations are taken from median. T/F

(vii) Lorenz curve was used for the first time for measuring the distribution of profits. T/F

(viii) Average deviation is most widely used in statistical work. T/F

(ix) In a moderately asymmetrical distribution, QD < MD < SD. T/F

(x) The variance is equal to square of standard deviation. T/F

Ans. (i) F, (ii) T, (iii) T, (iv) F, (v) F, (vi) T, (vii) F, (viii) F, (ix) T, (x) F

50. Find the mode, median, lower quartile (Q_1) and upper quartile (Q_3) and Coeff. of Q.D. from the following data:

Wages :	0–10	10–20	20–30	30–40	40–50
No. of workers :	22	38	46	35	20

$$\begin{bmatrix} \text{Mode} = 24.21\text{: Med.} = 24.46, \\ Q_1 = 14.803\text{: } Q_3 = 24.21 \\ \text{Coeff. of Q.D} = 0.396 \end{bmatrix}$$

51. Life of Allwyn and Godrej refrigerators that are currently popular were observed to be the following in a survey:

Life of refrigerators (in months)	***No. of refrigerators***	
	Allwyn	***Godrej***
0–20	10	8
20–40	28	20
40–60	24	32
60–80	10	28
80–100	8	12

(a) Compute coefficients of variations for the life of Allwyn and Godrej refrigerators.

(b) Which model do you prefer? Give reasons.

52. A purchasing agent obtained samples of 60 watt bulbs from two companies. He had the samples tested in his own laboratory for lengths of life with the following results:

Length of Life (in hours)	*Samples From* Co. A	Co.B
1700 and under 1900	10	3
1900 and under 2100	16	40
2100 and under 2300	20	12
2300 and under 2500	8	3
2500 and under 2700	6	2

(i) Which co's, bulbs do you think are better terms of average life?

(ii) If prices of both the co's, are same which co's, bulbs would you buy and why?

53. (a) What do your understand by 'Dispersion'? What purpose does a measure of dispersion serve?

(b) Define mean deviation. How does it differ from standard deviation?

54. (a) In what way measures of variation supplement measures of central tendency? Explain.

(b) What is coefficient of variation? What purpose does it serve? Also distinguish between 'variance' and 'coefficient of variation'.

55. (a) "Measures of dispersion and central tendency are complementary to each other in highlighting the characteristics of a frequency distribution." Explain the statement giving suitable examples.

(b) Discuss various measures of dispersion. Explain each with the help os one illustration.

56. (a) What do you mean by dispersion? Sate one property of standard deviation which you consider to be the most important.

(b) Explain with suitable example the term variation. What purpose does a measure of variation serve? Comment on some of the well-known insures of variation along with their respective merits and demerits.

(c) Give the uses of standard deviation.

(d) Define dispersion. Discuss briefly the absolute and relative measures of dispersion.

57. (a) List the various methods of measuring dispersion.

(b) Describe the various methods of measuring variation along with their respective merits and demerits.

(c) Calculate the standard deviation from the marks obtained by 5 students.

Marks out of 25:	8	12	13	15	22

(d) Explain with suitable example the term 'variation'. Mention some common measures of variation and describe the one which you think is widely used.

(e) What is 'dispersion'? What are the various measures of dispersion? Explain them fully.

(f) Fill is the blanks:

If C.V. of distribution is 50, SD = 20 the $\overline{X}$ shall be....

58. Two samples of size 40 to 50 have the same mean 53, but different standard deviation of the combined sample of size 90.

59. For a group of 200 candidates, the mean and standard deviations were found the be 40 and 15. Later on, it was discovered that the score 43 was missed as 53. Find the correct mean and standard deviation after correction.